The New Manufacturing

THE NEW MANUFACTURING

A Harvard Business Review Paperback

Harvard Business Review paperback No. 90080

The *Harvard Business Review* articles in this collection are available as individual reprints. Discounts apply to quantity purchases. For information and ordering contact Operations Department, Harvard Business School Publishing Division, Boston, MA 02163. Telephone: (617) 495-6192, 9 a.m. to 5 p.m. EST. Fax: (617) 495-6985, 24 hours a day.

Editor's Note: Some articles in this book may have been written before authors and editors began to take into consideration the role of women in management. We hope the archaic usage representing all managers as male does not detract from the usefulness of the collection.

Printed in the United States of America.
93 92 91 5 4 3 2 1

Contents

Strategies

Competing Globally

Stories from the Front Line

Make or Buy?

Collaborative Design

Strategies

Four concepts show us how the factory of 1999 must be built and managed.

The Emerging Theory of Manufacturing

by Peter F. Drucker

Japanese plants turn out two or three times more cars per worker than U.S. or European plants.

We cannot build it yet. But already we can specify the "postmodern" factory of 1999. Its essence will not be mechanical, though there will be plenty of machines. Its essence will be conceptual – the product of four principles and practices that together constitute a new approach to manufacturing.

Each of these concepts is being developed separately, by different people with different starting points and different agendas. Each concept has its own objectives and its own kinds of impact. Statistical Quality Control is changing the social organization of the factory. The new manufacturing accounting lets us make production decisions as business decisions. The "flotilla," or module, organization of the manufacturing process promises to combine the advantages of standardization and flexibility. Finally, the systems approach embeds the physical process of making things, that is, manufacturing, in the economic process of business, that is, the business of creating value.

As these four concepts develop, they are transforming how we think about manufacturing and how we manage it. Most manufacturing people in the United States now know we need a new theory of

Peter F. Drucker is the Clarke Professor of Social Science and Management at the Claremont Graduate School in Claremont, California. His most recent book is The New Realities *(Harper & Row, 1989).*

manufacturing. We know that patching up old theories has not worked and that further patching will only push us further behind. Together these concepts give us the foundation for the new theory we so badly need.

The most widely publicized of these concepts, Statistical Quality Control (SQC), is actually not new at all. It rests on statistical theory formulated 70 years ago by Sir Ronald Fisher. Walter Shewhart, a Bell Laboratories physicist, designed the original version of SQC in the 1930s for the zero-defects mass production of complex telephone exchanges and telephone sets. During World War II, W. Edwards Deming and Joseph Juran, both former members of Shewhart's circle, separately developed the versions used today.

The Japanese owe their leadership in manufacturing quality largely to their embrace of Deming's precepts in the 1950s and 1960s. Juran too had great impact in Japan. But U.S. industry ignored their contributions for 40 years and is only now converting to SQC, with companies such as Ford, General Motors, and Xerox among the new disciples. Western Europe also has largely ignored the concept. More important, even SQC's most successful practitioners do not thoroughly understand what it really does. Generally, it is considered a production tool. Actually, its greatest impact is on the factory's social organization.

By now, everyone with an interest in manufacturing knows that SQC is a rigorous, scientific method of identifying the quality and productivity that can be expected from a given production process in its current form so that control of both attributes can be built into the process itself. In addition, SQC can instantly spot malfunctions and show where they occur—a worn tool, a dirty spray gun, an overheating furnace. And because it can do this with a small sample, malfunctions are reported almost immediately, allowing machine operators to correct problems in real time. Further, SQC quickly identifies the impact of any change on the performance of the entire process. (Indeed, in some applications developed by Deming's Japanese disciples, computers can simulate the effects of a proposed change in advance.) Finally, SQC identifies where, and often how, the quality and productivity of the entire process can be continuously improved. This used to be called the "Shewhart Cycle" and then the "Deming Cycle"; now it is *kaizen*, the Japanese term for continuous improvement.

But these engineering characteristics explain only a fraction of SQC's results. Above all, they do not explain the productivity gap between Japanese and U.S. factories. Even after adjusting for their far greater reliance on outside suppliers, Toyota, Honda, and Nissan turn out two or three times more cars per worker than comparable U.S. or European plants do. Building quality into the process accounts for no more than one-third of this difference. Japan's major productivity gains are the result of social changes brought about by SQC.

The Japanese employ proportionately more machine operators in direct production work than Ford or GM. In fact, the introduction of SQC almost always increases the number of machine operators. But this increase is offset many times over by the sharp drop in the number of nonoperators: inspectors, above all, but also the people who do not *do* but *fix*, like repair crews and "fire fighters" of all kinds.

DRAWINGS BY WALLOP MANYUM

In U.S. factories, especially mass-production plants, such nonoperating, blue-collar employees substantially outnumber operators. In some plants, the ratio is two to one. Few of these workers are needed under SQC. Moreover, first-line supervisors also are gradually eliminated, with only a handful of trainers taking their place. In other words, not only does SQC make it possible for machine operators to be in control of their work, it makes such control almost mandatory. No one else has the hands-on knowledge needed to act effectively on the information that SQC constantly feeds back.

By aligning information with accountability, SQC resolves a heretofore irresolvable conflict. For more than a century, two basic approaches to manufacturing have prevailed, especially in the United States. One is the engineering approach pioneered by Frederick Winslow Taylor's "scientific management." The other is the "human relations" (or "human resources") approach developed before World War I by Andrew Carnegie, Julius Rosenwald of Sears Roebuck, and Hugo Münsterberg, a Harvard psychologist. The two approaches have always been considered antitheses, indeed, mutually exclusive. In SQC, they come together.

Taylor and his disciples were just as determined as Deming to build quality and productivity into the manufacturing process. Taylor asserted that his "one right way" guaranteed zero-defects quality; he was as vehemently opposed to inspectors as Deming is today. So was Henry Ford, who claimed that his assembly line built quality and productivity into the process (though he was otherwise untouched by Taylor's scientific management and probably did not even know about it). But without SQC's rigorous methodology, neither scientific management nor the assembly line could actually deliver built-in process control. With all their successes, both scientific management and the assembly line had to fall back on massive inspection, to fix problems rather than eliminate them.

> Most U.S. quality circles of the last 20 years have failed because they lacked rigorous and reliable feedback.

The human-relations approach sees the knowledge and pride of line workers as the greatest resource for controlling and improving quality and productivity. It too has had important successes. But without the kind of information SQC provides, you cannot readily distinguish productive activity from busyness. It is also hard to tell whether a proposed modification will truly improve the process or simply make things look better in one corner, only to make them worse overall.

Quality circles, which were actually invented and widely used in U.S. industry during World War II, have been successful in Japan because they came in after SQC had been established. As a result, both the quality circle and management have objective information about the effects of workers' suggestions. In contrast, most U.S. quality circles of the last 20 years have failed despite great enthusiasm, especially on the part of workers. The reason? They were established without SQC, that is, without rigorous and reliable feedback.

A good many U.S. manufacturers have built quality and productivity into their manufacturing processes without SQC and yet with a minimum of inspection and fixing. Johnson & Johnson is one such example. Other companies have successfully put machine operators in control of the manufacturing process without instituting SQC. IBM long ago replaced all first-line supervisors with a handful of "managers" whose main task is to train, while Herman Miller achieves zero-defects quality and high productivity through continuous training and productivity-sharing incentives.

But these are exceptions. In the main, the United States has lacked the methodology to build quality and productivity into the manufacturing process. Similarly, we have lacked the methodology to move responsibility for the process and control of it to the machine operator, to put into practice what the mathematician Norbert Wiener called the "human use of human beings."

SQC makes it possible to attain both traditional aspirations: high quality and productivity on the one hand, work worthy of human beings on the other. By fulfilling the aims of the traditional factory, it provides the capstone for the edifice of twentieth century manufacturing that Frederick Taylor and Henry Ford designed.

Bean counters do not enjoy a good press these days. They are blamed for all the ills that afflict U.S. manufacturing. But the bean counters will have the last laugh. In the factory of 1999, manufacturing accounting will play as big a role as it ever did and probably even a bigger one. But the beans will be counted differently. The new manufacturing accounting, which might more accurately be called "manufacturing economics," differs radically from traditional cost accounting in its basic concepts. Its aim is to integrate manufacturing with business strategy.

Manufacturing cost accounting (cost accounting's rarely used full name) is the third leg of the stool – the other legs being scientific management and the assembly line – on which modern manufacturing industry rests. Without cost accounting, these two could never have become fully effective. It too is American in origin. Developed in the 1920s by General Motors, General Electric, and Western Electric (AT&T's manufacturing arm), the new cost accounting, not technology, gave GM and GE the competitive edge that made them worldwide industry leaders. Following World War II, cost accounting became a major U.S. export.

But by that time, cost accounting's limitations also were becoming apparent. Four are particularly important. First, cost accounting is based on the realities of the 1920s, when direct, blue-collar labor accounted for 80% of all manufacturing costs other than raw materials. Consequently, cost accounting equates "cost" with direct labor costs. Everything else is "miscellaneous," lumped together as overhead.

These days, however, a plant in which direct labor costs run as high as 25% is a rare exception. Even in automobiles, the most labor intensive of the major industries, direct labor costs in up-to-date plants (such as those the Japanese are building in the United States and some of the new Ford plants) are down to 18%. And 8% to 12% is fast becoming the industrial norm. One large manufacturing company with a labor-intensive process, Beckman Instruments, now considers labor costs "miscellaneous." But typically, cost accounting systems are still based on direct labor costs that are carefully, indeed minutely, accounted for. The remaining costs – and that can mean 80% to 90% – are allocated by ratios that everyone knows are purely arbitrary and totally misleading: in direct proportion to a product's labor costs, for example, or to its dollar volume.

Second, the benefits of a change in process or in method are primarily defined in terms of labor cost savings. If other savings are considered at all, it is usually on the basis of the same arbitrary allocation by which costs other than direct labor are accounted for.

Even more serious is the third limitation, one built into the traditional cost accounting system. Like a sundial, which shows the hours when the sun shines but gives no information on a cloudy day or at night, traditional cost accounting measures only the costs of producing. It ignores the costs of nonproducing, whether they result from machine downtime or from quality defects that require scrapping or reworking a product or part.

Standard cost accounting assumes that the manufacturing process turns out good products 80% of the time. But we now know that even with the best SQC, nonproducing time consumes far more than 20% of total production time. In some plants, it accounts for 50%. And nonproducing time costs as much as producing time does – in wages, heat, lighting, interest, salaries, even raw materials. Yet the traditional system measures none of this.

Finally, manufacturing cost accounting assumes the factory is an isolated entity. Cost savings in the factory are "real." The rest is "speculation" – for example, the impact of a manufacturing process change on a product's acceptance in the market or on service quality. GM's plight since the 1970s illustrates the problem with this assumption. Marketing people were unhappy with top management's decision to build all car models, from Chevrolet to Cadillac, from the same small number of bodies, frames, and engines. But the cost accounting model showed that such commonality would produce substantial labor cost savings. And so marketing's argument that GM cars would lose customer appeal as they looked more and more alike was brushed aside as speculation. In effect, traditional cost accounting can hardly justify a product *improvement*, let alone a product or process *innovation*. Automation, for instance, shows up as a cost but almost never as a benefit.

> Traditional cost accounting can hardly justify a product *improvement*, let alone an *innovation*.

All this we have known for close to 40 years. And for 30 years, accounting scholars, government accountants, industry accountants, and accounting firms have worked hard to reform the system. They have made substantial improvements. But since the reform attempts tried to build on the traditional system, the original limitations remain.

What triggered the change to the new manufacturing accounting was the frustration of factory-automation equipment makers. The potential users, the people in the plants, badly wanted the new equipment. But top management could not be persuaded to spend the money on numerically controlled machine tools or robots that could rapidly change tools, fixtures, and molds. The benefits of automated equipment, we now know, lie primarily in the reduction of nonproducing time by improving quality (that is, getting it right the first time) and by sharply curtailing machine downtime in changing over from one model or product to another. But these gains cost accounting does not document.

Out of this frustration came Computer-Aided Manufacturing-International, or CAM-I, a cooperative effort by automation producers, multinational manufacturers, and accountants to develop a new cost accounting system. Started in 1986, CAM-I is just beginning to influence manufacturing practice. But already it has unleashed an intellectual revolution. The most exciting and innovative work in management today is found in accounting theory, with new concepts, new approaches, new methodology–even what might be called new economic philosophy–rapidly taking shape. And while there is enormous controversy over specifics, the lineaments of the new manufacturing accounting are becoming clearer every day.

As soon as CAM-I began its work, it became apparent that the traditional accounting system could not be reformed. It had to be replaced. Labor costs are clearly the wrong unit of measurement in manufacturing. But–and this is a new insight–so are all the other elements of production. The new measurement unit has to be time. The costs for a given period of time must be assumed to be fixed; there are no "variable" costs. Even material costs are more fixed than variable, since defective output uses as much material as good output does. The only thing that is both variable and controllable is how much time a given process takes. And "benefit" is whatever reduces that time. In one fell swoop, this insight eliminates the first three of cost accounting's four traditional limitations.

> In the new accounting, finished-goods inventory is a sunk cost, not an asset.

But the new cost concepts go even further by redefining what costs and benefits really are. For example, in the traditional cost accounting system, finished-goods inventory costs nothing because it does not absorb any direct labor. It is treated as an "asset." In the new manufacturing accounting, however, inventory of finished goods is a "sunk cost" (an economist's, not an accountant's, term). Stuff that sits in inventory does not earn anything. In fact, it ties down expensive money and absorbs time. As a result, its time costs are high. The new accounting measures these time costs against the benefits of finished-goods inventory (quicker customer service, for instance).

Yet manufacturing accounting still faces the challenge of eliminating the fourth limitation of traditional cost accounting: its inability to bring into the measurement of factory performance the impact of manufacturing changes on the total business–the return in the marketplace of an investment in automation, for instance, or the risk in not making an investment that would speed up production changeovers. The in-plant costs and benefits of such decisions can now be worked out with considerable accuracy. But the business consequences are indeed speculative. One can only say, "Surely, this should help us get more sales," or "If we don't do this, we risk falling behind in customer service." But how do you quantify such opinions?

Cost accounting's strength has always been that it confines itself to the measurable and thus gives objective answers. But if intangibles are brought into its equations, cost accounting will only raise more questions. How to proceed is thus hotly debated, and with good reason. Still, everyone agrees that these business impacts have to be integrated into the measurement of factory performance, that is, into manufacturing accounting. One way or another, the new accounting will force managers, both inside and outside the plant, to make manufacturing decisions as *business* decisions.

Henry Ford's epigram, "The customer can have any color as long as it's black," has entered American folklore. But few people realize what Ford meant: flexibility costs time and money, and the customer won't pay for it. Even fewer people realize that in the mid-1920s, the "new" cost accounting made it possible for GM to beat Ford by giving customers both colors and annual model changes at no additional cost.

By now, most manufacturers can do what GM learned to do roughly 70 years ago. Indeed, many go quite a bit further in combining standardization with flexibility. They can, for example, build a variety of end products from a fairly small number of standardized parts. Still, manufacturing people tend to think like Henry Ford: you can have either standardization at low cost or flexibility at high cost, but not both.

The factory of 1999, however, will be based on the premise that you not only *can* have both but also *must* have both–and at low cost. But to achieve this, the factory will have to be structured quite differently.

Today's factory is a battleship. The plant of 1999 will be a "flotilla," consisting of modules centered either around a stage in the production process or around a number of closely related operations. Though overall command and control will still exist, each module will have its own

command and control. And each, like the ships in a flotilla, will be maneuverable, both in terms of its position in the entire process and its relationship to other modules. This organization will give each module the benefits of standardization and, at the same time, give the whole process greater flexibility. Thus it will allow rapid changes in design and product, rapid response to market demands, and low-cost production of "options" or "specials" in fairly small batches.

No such plant exists today. No one can yet build it. But many manufacturers, large and small, are moving toward the flotilla structure: among them are some of Westinghouse's U.S. plants, Asea Brown Boveri's robotics plant in Sweden, and several large printing plants, especially in Japan.

The biggest impetus for this development probably came from GM's failure to get a return on its massive (at least $30 billion and perhaps $40 billion) investment in automation. GM, it seems, used the new machines to improve its existing process, that is, to make the assembly line more efficient. But the process instead became less flexible and less able to accomplish rapid change.

Meanwhile, Japanese automakers and Ford were spending less and attaining more flexibility. In these plants, the line still exists, but it is discontinuous rather than tightly tied together. The new equipment is being used to speed changes, for example, automating changeovers of jigs, tools, and fixtures. So the line has acquired a good bit of the flexibility of traditional batch production without losing its standardization. Standardization and flexibility are thus no longer an either-or proposition. They are – as indeed they must be – melded together.

This means a different balance between standardization and flexibility, however, for different parts of the manufacturing process. An "average" balance across the plant will do nothing very well. If imposed throughout the line, it will simply result in high rigidity and big costs for the entire process, which is apparently what happened at GM. What is required is a reorganization of the process into modules, each with its own optimal balance.

Moreover, the relationships between these modules may have to change whenever the product, process, or distribution changes. Switching from selling heavy equipment to leasing it, for instance, may drastically change the ratio between finished-product output and spare-parts output. Or a fairly minor model change may alter the sequence in which major parts are assembled into the finished product. There is nothing very new in this, of course. But under the traditional line structure, such changes are ignored, or they take forever to accomplish. With competition intensifying and product life cycles shortening all the time, such changes cannot be ignored, and they have to be done fast. Hence the flotilla's modular organization.

But this organization requires more than a fairly drastic change in the factory's physical structure. It requires, above all, different communication and information. In the traditional plant, each sector and department reports separately upstairs. And it reports what upstairs has asked for. In the factory of 1999, sectors and departments will have to think through what information they owe to whom and what information they need from whom. A good deal of this information will flow sideways and across department lines, not upstairs. The factory of 1999 will be an information network.

Consequently, all the managers in a plant will have to know and understand the entire process, just as the destroyer commander has to know and understand the tactical plan of the entire flotilla. In the factory of 1999, managers will have to

Today's factory is a battleship. Tomorrow's will be a flotilla.

think and act as team members, mindful of the performance of the whole. Above all, they will have to ask: What do the people running the other modules need to know about the characteristics, the capacity, the plans, and the performance of *my* unit? And what, in turn, do we in my module need to know about theirs?

The last of the new concepts transforming manufacturing is systems design, in which the whole of manufacturing is seen as an integrated process that converts materials into goods, that is, into economic satisfactions.

Marks & Spencer, the British retail chain, designed the first such system in the 1930s. Marks & Spencer designs and tests the goods (whether textiles or foods) it has decided to sell. It designates one manufacturer to make each product under contract. It works with the manufacturer to produce the right merchandise with the right quality at the right price. Finally, it organizes just-in-time delivery of the finished products to its stores. The entire process is governed by a meticulous forecast as to when the goods will move off store shelves and into customers' shopping bags. In the last ten years or so, such systems management has become common in retailing.

Henry Ford did not build a system at River Rouge. He built an unwieldy monster.

Though systems organization is still rare in manufacturing, it was actually first attempted there. In the early 1920s, when the Model T was in its full glory, Henry Ford decided to control the entire process of making and moving all the supplies and parts needed by his new plant, the gigantic River Rouge. He built his own steel mill and glass plant. He founded plantations in Brazil to grow rubber for tires. He bought the railroad that brought supplies to River Rouge and carried away the finished cars. He even toyed with the idea of building his own service centers nationwide and staffing them with mechanics trained in Ford-owned schools. But Ford conceived of all this as a financial edifice held together by ownership. Instead of building a system, he built a conglomerate, an unwieldy monster that was expensive, unmanageable, and horrendously unprofitable.

In contrast, the new manufacturing system is not "controlled" at all. Most of its parts are independent – independent suppliers at one end, customers at the other. Nor is it plant centered, as Ford's organization was. The new system sees the plant as little more than a wide place in the manufacturing stream. Planning and scheduling start with shipment to the final customer, just as they do at Marks & Spencer. Delays, halts, and redundancies have to be designed into the system – a warehouse here, an extra supply of parts and tools there, a stock of old products that are no longer being made but are still occasionally demanded by the market. These are necessary imperfections in a continuous flow that is governed and directed by information.

What has pushed American manufacturers into such systems design is the trouble they encountered when they copied Japan's just-in-time methods for supplying plants with materials and parts. The trouble could have been predicted, for the Japanese scheme is founded in social and logistic conditions unique to that country and unknown in the United States. Yet the shift seemed to American manufacturers a matter of procedure, indeed, almost trivial. Company after company found, however, that just-in-time delivery of supplies and parts created turbulence throughout their plants. And while no one could figure out what the problem was, the one thing that became clear was that with just-in-time deliveries, the plant no longer functions as a step-by-step process that begins at the receiving dock and ends when finished goods move into the shipping room. Instead, the plant must be redesigned from the end backwards and managed as an integrated flow.

Manufacturing experts, executives, and professors have urged such an approach for two or three decades now. And some industries, such as petroleum refining and large-scale construction, do practice it. But by and large, American and European manufacturing plants are neither systems designed nor systems managed. In fact, few companies have enough knowledge about what goes on in their plants to run them as systems. Just-in-time delivery, however, forces managers to ask systems questions: Where in the plant do we need redundancy? Where should we place the burden of adjustments? What costs should we incur in one place to minimize delay, risk, and vulnerability in another?

Producing does not stop when the product leaves the factory. Distribution and service are integral parts of the process.

A few companies are even beginning to extend the systems concept of manufacturing beyond the plant and into the marketplace. Caterpillar, for instance, organizes its manufacturing to supply any replacement part anywhere in the world within 48 hours. But companies like this are still exceptions; they must become the rule. As soon as we define manufacturing as the process that converts things into economic satisfactions, it becomes clear that producing does not stop when the product leaves the factory. Physical distribution and product service are still part of the production process and should be integrated with it, coordinated with it, managed together with it. It is already widely recognized that servicing the product must be a major consideration during its design and production. By 1999, systems manufacturing will have an increasing influence on how we design and remodel plants and on how we manage manufacturing businesses.

Traditionally, manufacturing businesses have been organized "in series," with functions such as engineering, manufacturing, and marketing as successive steps. These days, that system is often complemented by a parallel team organization (Procter & Gamble's product management teams are a well-known example), which brings various functions together from the inception of a new product or process project. If manufacturing is a system, however, every decision in a manufacturing business becomes a manufacturing decision. Every decision should meet manufacturing's requirements and needs and in turn should exploit the strengths and capabilities of a company's particular manufacturing system.

When Honda decided six or seven years ago to make a new, upscale car for the U.S. market, the most heated strategic debate was not about design, performance, or price. It was about whether to distribute the Acura through Honda's well-established dealer network or to create a new market segment by building separate Acura dealerships at high cost and risk. This was a marketing issue, of course. But the decision was made by a team of design, engineering, manufacturing, and marketing people. And what tilted the balance toward the separate dealer network was a manufacturing consideration: the design for which independent distribution and service made most sense was the design that best utilized Honda's manufacturing capabilities.

Full realization of the systems concept in manufacturing is years away. It may not require a new Henry Ford. But it will certainly require very different management and very different managers. Every manager in tomorrow's manufacturing business will have to know and understand the manufacturing system. We might well adopt the Japanese custom of starting all new management people in the plant and in manufacturing jobs for the first few years of their careers. Indeed, we might go even further and require managers throughout the company to rotate into factory assignments throughout their careers–just as army officers return regularly to troop duty.

In the new manufacturing business, manufacturing is the integrator that ties everything together. It creates the economic value that pays for everything and everybody. Thus the greatest impact of the manufacturing systems concept will not be on the production process. As with SQC, its greatest impact will be on social and human concerns–on career ladders, for instance, or more important, on the transformation of *functional* managers into *business* managers, each with a specific role, but all members of the same production and the same cast. And surely, the manufacturing businesses of tomorrow will not be run by financial executives, marketers, or lawyers inexperienced in manufacturing, as so many U.S. companies are today.

There are important differences among these four concepts. Consider, for instance, what each means by "the factory." In SQC, the factory is a place where people work. In management accounting and the flotilla concept of flexible manufacturing, it is a place where work is being done–it makes no difference whether by people, by white mice, or

Caterpillar organizes its manufacturing to supply any part anywhere in the world within 48 hours.

by robots. In the systems concept, the factory is not a place at all; it is a stage in a process that adds economic value to materials. In theory, at least, the factory cannot and certainly should not be designed, let alone built, until the entire process of "making"–

Understand the "making" process all the way to the final customer. Then design and build the factory.

all the way to the final customer–is understood. Thus defining the factory is much more than a theoretical or semantic exercise. It has immediate practical consequences on plant design, location, and size; on what activities are to be brought together in one manufacturing complex; even on how much and in what to invest.

Similarly, each of these concepts reflects a particular mind-set. To apply SQC, you don't have to think, you have to do. Management accounting concentrates on technical analysis, while the flotilla concept focuses on organization design and work flow. In the systems concept, there is great temptation to keep on thinking and never get to the doing. Each concept has its own tools, its own language, and addresses different people.

Nevertheless, what these four concepts have in common is far more important than their differences. Nowhere is this more apparent than in their assumption that the manufacturing process is a configuration, a whole that is greater than the sum of its parts. Traditional approaches all see the factory as a collection of individual machines and individual operations. The nineteenth century factory was an assemblage of machines. Taylor's scientific management broke up each job into individual operations and then put those operations together into new and different jobs. "Modern" twentieth century concepts–the assembly line and cost accounting–define performance as the sum of lowest cost operations. But none of the new concepts is much concerned with performance of the parts. Indeed, the parts as such can only underperform. The process produces results.

Management also will reflect this new perspective. SQC is the most nearly conventional in its implications for managers, since it does not so much change their job as shift much of it to the work force. But even managers with no business responsibility (and under SQC, plant people have none) will have to manage with an awareness of business considerations well beyond the plant. And every manufacturing manager will be responsible for integrating people, materials, machines, and time. Thus every manufacturing manager ten years hence will have to learn and practice a discipline that integrates engineering, management of people, and business economics into the manufacturing process. Quite a few manufacturing people are doing this already, of course–though usually unaware that they are doing something new and different. Yet such a discipline has not been systematized and is still not taught in engineering schools or business schools.

These four concepts are synergistic in the best sense of this much-abused term. Together–but only together–they tackle the conflicts that have most troubled traditional, twentieth century mass-production plants: the conflicts between people and machines, time and money, standardization and flexibility, and functions and systems. The key is that every one of these concepts defines performance as productivity and conceives of manufacturing as the physical process that adds economic value to materials. Each tries to provide economic value in a different way. But they share the same theory of manufacturing.

Author's note: I wish to acknowledge gratefully the advice and criticism I received on this piece from Bela Gold and Joseph Maciariello, friends and colleagues at the Claremont Graduate School.

Reprint 90303

Elizabeth A. Haas

Breakthrough manufacturing

Is the decay of American manufacturing really incurable? Massive capital investment has failed to check it. The $200 billion that U.S. companies poured into new domestic facilities and equipment between 1980 and 1985 scarcely slowed the erosion of our global competitiveness. The U.S. auto industry, which alone invested $40 billion in capital improvements in that period, continues to lose ground. Six years ago, Japan's automakers were turning out subcompacts at a cost $1,500 less than Detroit's; today the gap has widened to $1,800. Nor is cost our only problem. Last year, American-made cars were still three times as likely as their Japanese counterparts to need repair in the first 12 months of ownership. Comparable cost and quality problems continue to plague nearly every corner of U.S. manufacturing industry.

All manufacturing decisions must now meet the test of fierce global competition.

It is my conviction that the world competitiveness of U.S. industry could be lost for good, well before the end of this century, unless American top managers can learn to think and act in a radically different way in their everyday manufacturing decisions.

By and large, U.S. manufacturers are still the slaves of operational necessity. When need dictates–when a new product is introduced, when volume fluctuates, or when costs have to be cut–they make operational changes. Over time, these changes may marginally enhance a company's competitiveness or at least slow its erosion. Rarely, if ever, do they add up to a coherent strategy for gaining an edge over the competition.

The difference between an operational and a strategic perspective is immense. It is the difference between a focus on operational improvement–"How can we do it better than before?"–and a focus on strategic advantage–"How can we use it to beat the competition?" Managers who are preoccupied with optimizing the operational bits and pieces–reducing machining time here, paring labor costs there–will typically be blind to the big strategic opportunities. It is only by managing the operational components as an integrated system that manufacturers can exploit their full potential for delivering added value to customers through lower prices, greater service responsiveness, or higher quality.

Typically, manufacturing decisions are still taken in an operational framework defined by internal performance standards–machine downtime, scrap rate, work-in-process inventories, and the like. But the real test of manufacturing decisions is their impact on the company's performance in the dog-eat-dog world of global competition. Strategic decision making aims to put the company ahead of its competitors and keep it there through positive differentiation–sometimes in defiance of operational logic.

A manufacturing decision that might be downright stupid in operational terms alone may look very different when seen from a strategic perspective. Consider a simple example. An appliance maker found that it could extend the life of its product and save $1.75 per unit in warranty costs by substituting a newly designed part. Against expectation, it turned out that the new part would add more than $2 to the cost of each unit–hardly an attractive proposition, operationally speaking. But on strategic grounds the move still made excellent sense. By sparing the end user cost and inconvenience over the product's lifetime, it would strengthen the company's sales message and increase the product's value in the customer's eye. Management's decision to go ahead has since been rewarded by a hefty gain in market share.

Elizabeth Haas, a principal of McKinsey & Company, works with the senior management of client companies on manufacturing strategy and advanced technology issues. She has written for the Wall Street Journal, *the* Operations Research Journal of America, *and other professional publications.*

The strategic breakpoint

How much must a company differentiate its products to gain a decisive competitive advantage? Experience indicates that there is a definable point at which the market will respond to a change in value–a point where an incremental improvement in some value parameter (price, quality, or service) will trigger a disproportionate volume increase and tilt the competitive balance. We call this the strategic breakpoint.

The success of Japanese cars in the U.S. auto market is a case in point. When Toyotas had to go in for repair every eight weeks on average and American cars every six weeks, few buyers noticed the difference. But when Toyotas began to average six months between repairs and American cars seven weeks, the gap was too big to miss. Using noticeably higher quality, at competitive prices, as a lever to gain market share, the Japanese had reached a strategic breakpoint. All at once, it seemed customers en masse were choosing Japan's product over Detroit's.

Efficient operations aren't enough to gain competitive advantage.

Kennametal Inc., a U.S. carbide tool manufacturer, found its strategic breakpoint by improving the quality of its products. Aiming for a clear competitive advantage, Quentin McKenna, the chief executive, decreed a new quality standard of zero defects, backed by a guarantee to the customer that every Kennametal tool would be perfect on delivery. Making good on that promise meant changing to a costlier production process, but the result enabled Kennametal's customers to use automatic tool-changing devices on their equipment and dispense with prior inspection. Six months after McKenna set the new standard, Kennametal had doubled its market share.

A U.S. printed-circuit-board fabricator achieved a strategic breakpoint by adopting an innovative production process that looked uneconomic in operational terms. Wanting to cut costs, management had asked its engineers to experiment with a new chemical deposition process for making printed-circuit boards. When the engineering team had finally worked out all the bugs, management learned that the new technology would add 30% to the boards' cost. But it would also produce denser boards of better quality that could be sold at high margins in a specialty market. An operationally unattractive process turned out to make superb strategic sense.

U.S. manufacturers have habitually thought of cost, quality, and service in terms of trade-offs. They have taken it for granted that higher quality means higher manufacturing cost or longer throughput time. Sometimes it does take that kind of trade-off to achieve a strategic breakpoint, but it usually does not. Quite often, it proves possible to achieve substantial improvement along two or three of these dimensions at once.

Critical decisions

Broadly speaking, there are eight kinds of manufacturing decisions: product design, process design, facility and plant configuration, information and control systems, human resources, research and development, suppliers' roles and relationships, and organization. Changes in any of these areas can initially affect some or all of the others, and understanding their interrelationships is basic to achieving a strategic breakpoint.

Product design. A primary driver of cost, quality, and factory throughput is product design. Shrewdly leveraged, the design process can lead to a strategic breakpoint–one that may even alter the company's entire business strategy. Faced with rising competition from the Far East, one American consumer electronics manufacturer found that moving production offshore could save 10% to 15% on costs. Not only did suppliers in Southeast Asia have the necessary engineering expertise, but also components, labor, and capital would be cheaper and delivery times shorter. But recent advances in the material properties of plastics, which happened to be 20% cheaper in the United States than in the Far East, offered management an opportunity to reduce the cost of materials and assembly by radically redesigning the product. Today, the company makes its product in the United States at a cost lower than any of its competitors.'

A supplier of cold-formed metal parts also used product design to attain a strategic breakpoint. To benefit from joint learning and group manufacturing, this supplier shifted from a strategy of supplying a broad range of products to a few targeted customers to a strategy of marketing a limited line of products with comparable geometric shapes and manufacturing requirements to a much broader market segment. Fewer geometric shapes meant quicker production-line changeovers from one product to another, and the structure of the product line meant the workers were more familiar with the task at hand. The company nearly doubled its return on sales and increased its profits by 30% last year.

Product design can affect parts and tooling requirements, dictate tasks for the work force, constrain or facilitate new processes, and pose special scheduling or quality-control problems. Thus, even when it is not the fulcrum of manufacturing strategy, design still plays a crucial role.

Process design. Not long ago, decisions on manufacturing processes, equipment, and tooling were routine. New product introductions and the need to replace worn-out equipment dominated investment decisions. Except for incremental changes in equipment and systems, process innovation was rarely at the top of management's agenda.

All that is changing. Today, manufacturing executives are more alert to the strategic potential of process design. Recently, a computer components manufacturer invested $20 million in a flexible assembly system. The investment made good operational sense because it paid for itself in less than a year. Strategically, the investment was even more attractive. Production time was cut by 80%, and product quality (measured as a percentage of units "dead on arrival") improved tenfold. The formerly undistinguished company is now seen as an outstanding U.S. manufacturer.

Significant process-design decisions don't necessarily require massive investments in new equipment and systems. An electronic equipment company had planned to build a "factory of the future" by replacing three old production lines with five automated lines. When management realized that the new process wouldn't necessarily ensure the same level of quality, it decided to think the whole project through again. In the end, it decided on a more conservative course–selectively upgrading and rearranging its existing machines and adding materials-handling devices. By making these relatively minor changes in its existing plants, the company improved both throughput and quality to the point where the three renovated lines could handle the anticipated increase in volume. Customers gained through lower prices for higher quality goods.

Many companies invest heavily in process design without getting the benefits they want. They neglect to integrate changes in the other seven manufacturing areas or they fail to set clear strategic objectives. If the organizational groundwork isn't properly laid, for example, process-design changes may have little lasting effect. Workers on the shop floor may drift back to their old ways of dealing with production problems. Maintenance workers may delay preventive maintenance because of time pressures created by poor organizational planning. Purchasing agents, still preoccupied with cutting costs, may undermine quality by ordering cheaper but less uniform materials. These slippages can cause the deterioration of the entire production system.

Facility and plant configuration. In larger companies, improved manufacturing techniques often cause excess capacity, making reconfigurations and painful plant closings inevitable. But when implemented as the first step in a manufacturing strategy rather than later, after other decisions are made, reconfiguration can be a valuable catalyst for change. Because it compels management to take a fresh look at accepted practice, reconfiguration can trigger necessary changes in other areas.

Consider the situation confronting Reliance Electric Company eight years ago. In a push for domestic market leadership, management launched a drive to cut costs and boost operating efficiency in its mechanical group. But no sooner had the company achieved that goal than it found itself looking at dangerous new competition from the Far East.

A young engineer suggested a drastic change in plant layout. Instead of the existing arrangement–a process-oriented setup that grouped similar machines together, requiring products to be moved in batches from one station to the next–he recommended a layout based on product-specific manufacturing cells so that workers could do a complete sequence of operations on a single product. Convinced that this idea made sense, management moved production from two older plants to three new, smaller facilities to ensure that outmoded shop-floor practices wouldn't hamper production on the new lines. Management also ordered new tools and rearranged equipment. The result: Reliance not only withstood the competitive threat from offshore producers but emerged with the dominant market share among its domestic competitors.

Information and control systems. The bewildering variety of powerful systems that can be applied to factory-floor operations often lure top managers into a trap: they invest in new systems before thinking through their strategic objectives. Systems development priorities and resource allocation ought always to be guided by the strategic breakpoint that management aims to achieve. If greater responsiveness to customers is the goal, then systems to eliminate unpredictable elements in processes and thereby support part and product flows should probably head the agenda. A quality breakpoint, on the other hand, may call for developing shop-floor controls and feedback systems first.

When Hewlett-Packard's management asked why quality-control engineers needed almost a month to diagnose failures in the company's hard-disk packs, it learned that for each pack, data had to be collected on more than 400 quality-control measures. So

"The Japanese have already surpassed us in compact discs and I don't even know what they are."

numerous were the possible causes of pack failure and so complex were their interrelationships that the engineers often had to spend three to four weeks finding the cause of each failure.

By installing a computerized system to track and interpret the quality-control data, Hewlett-Packard reduced both the failure rate and the time required to correct defects by more than 90%. In addition, materials and labor costs were halved, and the time needed to serve customers was reduced by two-thirds.

Since systems decisions are inextricably tied to other operational functions, the right systems decisions can alleviate bottlenecks, eliminate the unpredictability of a manufacturing process, improve product and part availability, and enhance customer service. Before management automates a process, for example, systems may have to be installed to monitor production stations or read bar codes of incoming parts. Similarly, it may not be possible to speed up product introductions or broaden the product line without first upgrading the computer-aided design and scheduling systems. In short, before other operational changes can be implemented, system development may be necessary.

Human resources. Companies seeking a cost advantage over competitors usually zero in early on labor cost. Wage levels in particular are an obvious target. Cutting wages and benefits may gain a temporary reprieve from Chapter 11, but they are rarely the key to a sustainable competitive advantage. Because they sour work-force attitudes, wage cuts often put out of reach the productivity gains that could be obtained by more skilled, sensitive management of human resources.

Training the work force in added skills, for example, can permit more flexibility in job assignments, and that in turn can boost shop-floor productivity. By paying a wage increment for each new skill learned, Sherwin-Williams gave employees at one of its automotive paint plants an incentive to broaden their job capabilities. A year after this program was launched, more than 95% of the work force had developed new skills. Two and a half years later, product quality had improved and labor costs were actually cut in half.

A European consumer goods manufacturer not only pared its costs but also improved its responsiveness to customers with an innovative human resource policy. The company's business was highly seasonal, requiring it to ship out more than 80% of its production in June and July. To deal with the seasonal surge in demand, management had historically relied on inventory buildup, overtime, and temporary hiring. Operational foul-ups were common. Because it had to build inventory far in advance of shipment, the company had trouble adjusting the product mix to shifts in customer demand. During the height of the selling season, management was plagued with schedule changes, interruptions, complaints of damaged or mishandled goods, and urgent calls from customers for parts.

To recapture its competitive edge, the company persuaded its workers to accept a novel contract. They would continue to be paid throughout the year on a 40-hour-a-week basis, but the scheduling of work would change. From March through mid-July, they would work a 50-hour week assembling and painting the products. From August through November, a low-demand period, they would work a 30-hour week producing components common to several models.

This more flexible use of labor has enabled the company to double its capacity in the peak season and to reduce its working capital requirement 40%. The company has become competitive in price while maintaining a quality advantage.

Despite their power, human resource decisions are risky. They impinge heavily on other decision areas, and their effects are often irreversible. Restructuring a company's human resource policy can have such a profound impact on attitudes and expectations that it probably should not be attempted more than once in a management generation. Hence management should carefully think through proposed changes in the other manufacturing areas before attempting to revamp its human resource policy.

Research and development. Because R&D is the key to developing proprietary process and product expertise, it influences almost all other manufacturing decisions, even those involving roles and relationships with suppliers (for example, what technology suppliers will provide), facility location, and development of special organizational capabilities.

Most people credit Polaroid Corporation's preeminence in instant photography to its founder's invention of the first instant camera in the mid-1950s. But Edwin Land's breakthrough doesn't explain why 30 years later Polaroid still reigns supreme in its market. The real reason has more to do with superior process technology than with product innovation. Having defined its unique product feature as "developing film in air," Polaroid has focused its research on continually improving its proprietary process for finishing spreader rollers – the component controlling picture quality. High-speed, low-cost coating and assembly of multilayered film and batteries, another unique skill perfected through patient R&D, gave Polaroid the opportunity to diversify into the coating of computer disks.

When management is alert to current developments in both products and process technology, it may take a very different approach to thinking about next-generation products and other manufacturing decisions. A Far East supplier offered to sell precision-cast metal turbine blades to a U.S. aircraft engine manufacturer for less than the company's current cost. Management hesitated, worrying that the arrangement might tempt the supplier to enter the engine business itself. But the research group came up with convincing evidence that metal blades would eventually be replaced by ceramic blades, which had the advantages of greater strength, lighter weight, and faster heat dissipation. Reassured, management decided to source its blades from the Far East and shift its R&D in the meantime from improving the casting process to developing processes for working with the next generation of materials.

Had its strategic planning process not forced management to look at the implications of basic materials science, this company would have forgone significant short-term savings and missed out on an important long-term technological opportunity.

Suppliers' roles and relationships. Lacking the in-house expertise needed to keep up with advances in all the relevant technologies, few companies can excel in every aspect of manufacturing. For most, it is uneconomic even to try. But there is an alternative.

Global competition is pushing suppliers to develop new specialized capabilities, and by tapping their technological expertise, some manufacturers have gained the flexibility needed to reach a strategic breakpoint.

Viewed strategically, decisions on suppliers' roles and relationships go beyond simple choices about what to make in-house. They are decisions about how to position the company's manufacturing capability so as to get the best value from suppliers while also maximizing the competitive value of in-house skills. And they are decisions about how to work with suppliers – by subcontracting, joint research contracts, or joint ventures, for example – to establish the best long-term working relationships. As a result, decisions about how to work with suppliers become tightly linked to process design and product design decisions, organizational decisions, and sometimes systems decisions as well – as, for example, when a company links its scheduling systems with those of its suppliers.

Manufacturing practices and strategies should try to tilt the competitive balance.

The U.S. auto industry is a good example of this trend. Parts suppliers, who used to bid only after receiving a request for a proposal, now help set vehicle performance objectives, design parts and processes, and lend auto manufacturers a hand in other aspects of product development. These developments have reduced the number of parts manufacturers by two-thirds while giving the automakers improved responsiveness and better product quality at less cost. Chrysler attributes a 50% reduction in engineering costs and the achievement of a quality breakpoint in product design largely to its greater reliance on suppliers.

In the highly cyclical electronics industry, manufacturers have habitually ordered from their suppliers on a very short-term basis. Hewlett-Packard took a different tack. It committed itself to firm orders six months ahead. This change enabled Hewlett-Packard's silicon suppliers to exploit economies of scale and improve their quality. In turn, Hewlett-Packard could rely on component quality and delivery even in times of industry shortage, thereby better serving its own customers and reaching a strategic breakpoint.

Organization. Rarely can a business achieve its strategic breakpoint through organizational change alone. By changing its organizational procedures, structures, skills, and style, however, management can relax operational constraints, foster coherent decision making, and enhance its efforts to reach a strategic breakpoint.

Take the case of a domestic automotive parts supplier. To design and manufacture a new product planned as a major entry in the world market, management needed to install complex, integrated automation and computer systems. These demanded more rigorous planning and control. Management saw that if the new technology was to be effectively deployed, organizational change would be necessary.

Instead of handing the assignment to technical experts on its central advisory staff, management set up a cross-functional project team and gave it full responsibility for both planning and implementation. Lodged in separate facilities, the team was given all the resources it needed and told to carry on without regard to established practices or historical loyalties. The organizational changes have yielded hundreds of improvements in integrating process and product development along with process breakthroughs and novel applications of old technology. Engineering and marketing now work hand in hand. The company is meeting rigorous new cost standards, maintaining a zero-defect quality standard, and reaching its operating targets (single-unit lot sizes).

When a building materials manufacturer's computer-aided design system didn't perform as promised, management scrapped it and assigned a task force assembled from marketing, R&D, and other functions to come up with a new one. Six months later, the task force had implemented a CAD system that not only improved materials manufacturing but also allowed the company to diversify into building design. This manufacturer now leads the industry in cost and customer responsiveness. Using a task force as a temporary bridging structure was the organizational change that helped it achieve a strategic breakpoint.

Integrated decisions

In today's environment, there are no unitary solutions. Whatever the strategic breakpoint management has targeted, it is never possible to isolate a single decision that would make everything happen. The most successful corporations not only take the interactions among the eight critical decision areas into account; they also continually reevaluate and reorchestrate their manufacturing decisions to support their strategic goals.

Companies often pay a high price for failing to make these decisions from a strategic perspective. Take what can happen with a decision to reconfigure facilities – an obvious option for managers wanting to serve their local markets better. To sharpen its response to seasonal demand, a clothing company moved its production from four giant plants to a dozen smaller facilities. By reducing the maximum distance from any marketing center to its plant from nearly 3,000 miles to about 300, the company management was confident that it could better serve the whims of regional markets. At the same time, it invested in advanced equipment to make its production faster and more flexible.

As the spring season approached, management ramped up production. Initial sales reports signaled a need to alter the product mix and production schedules, and unexpected bottlenecks developed. The company's scheduling, purchasing, and distribution systems proved unable to cope with the new mode of operation.

It's a mistake to think of manufacturing as just production.

That spring was a disaster. Hundreds of little things blocked change. For example, any alteration in a single plant's schedule automatically regenerated the schedules of all the other plants – a process that took 48 hours to complete. Even so, all the plant managers insisted on having the flexibility to alter their own production schedules. The result was an unpredictable sequence of daily, even hourly, changes. And although suppliers had agreed to keep on altering deliveries to accommodate schedule changes, they couldn't adjust swiftly enough to support the changes the company was now demanding.

The company's existing systems might have supported the new plant configuration, given a great deal of policy intervention, special rules for dealing with exceptions, and handling changes in the customary batch format. Even so, reliance on such piecemeal remedies would have been awkward, costly, and inherently vulnerable to failure.

In the course of the following year, as the systems were recast to support the new equipment, it became painfully obvious that the facility reconfiguration had been unnecessary. The capabilities of the new equipment, combined with the newly designed scheduling system had given the company far more flexibility than anyone had anticipated, and the reconfiguration had merely complicated the company's production logistics.

Western Electric (now AT&T Technologies) also encountered problems when it installed new computer-integrated equipment to remedy a quality problem that was dragging output down. Because management failed to integrate this change with other manufacturing decisions, a host of minor troubles frustrated implementation. When the equipment needed maintenance, workers hesitated to touch it for fear of distorting data in the system. When an atypical circuit board came down the line, workers would stop the line until they had checked the records to make sure it wasn't the result of a computer error. When materials ran short, the purchasing people hesitated to use the established rules for parts substitution because they weren't sure that the new system was compatible with the old procedures.

Success came only after Western Electric built a plant to house the new equipment and processes, established a full-time, high-level team to oversee the installation, and set about changing people's habits and assumptions. Incentives were introduced to encourage risk taking. Factory operators were trained to work with the computer system, and communication programs were launched to convert the organization to new ways of thinking about production. Western Electric's management learned that manufacturing changes cannot be made in isolation. To work effectively, the new computer-integrated equipment required a whole array of unanticipated adjustments, innovations, and leadership initiatives.

Properly linked, decisions in the eight manufacturing areas can deliver strategic breakpoints, but their interdependencies, if not taken into account, can also pose strategic barriers. Because these interrelationships are complex, what seems an obvious move may sometimes be neither feasible nor effective. At other times, an obvious move may be needlessly risky compared with a less dramatic alternative that could do the job with less disruption or at lower cost.

Technology's dramatic transformation of the factory has strengthened the link between manufacturing strategy and business strategy – and thereby invalidated a host of time-tested operational principles and decision criteria. More and more, competitive advantage will go to the companies that seek strategic breakpoints through the integration of decisions in every area of manufacturing.

Many American manufacturing companies cannot hope to be competitive unless they increase productivity by 10% to 15% annually during the next four to five years. Companies that move now to exploit the full power of manufacturing as a competitive weapon should find a commanding strategic advantage within their reach. But for those that continue to seek salvation in operational improvements, the long-term outlook is bleak.

Reprint 87205

"I have smarts, but they don't happen to be street smarts."

As a strategic weapon, time is the equivalent of money, productivity, quality, even innovation.

Time – The Next Source of Competitive Advantage

by GEORGE STALK, JR.

Like competition itself, competitive advantage is a constantly moving target. For any company in any industry, the key is not to get stuck with a single simple notion of its source of advantage. The best competitors, the most successful ones, know how to keep moving and always stay on the cutting edge.

Today, *time* is on the cutting edge. The ways leading companies manage time – in production, in new product development and introduction, in sales and distribution – represent the most powerful new sources of competitive advantage. Though certain Western companies are pursuing these advantages, Japanese experience and practice provide the most instructive examples – not because they are necessarily unique but because they best illustrate the evolutionary stages through which leading companies have advanced.

In the period immediately following World War II, Japanese companies used their low labor costs to gain entry to various industries. As wage rates rose and technology became more significant, the Japanese shifted first to scale-based strategies and then to focused factories to achieve advantage. The advent of just-in-time production brought with it a move to flexible factories, as leading Japanese companies sought both low cost and great variety in the market. Cutting-edge Japanese companies today are capitalizing on time as a critical source of competitive advantage: shortening the planning loop in the product development cycle and trimming process time in the factory – managing time the way most companies manage costs, quality, or inventory.

In fact, as a strategic weapon, time is the equivalent of money, productivity, quality, even innovation. Managing time has enabled top Japanese companies not only to reduce their costs but also to offer broad product lines, cover more market segments, and upgrade the technological sophistication of their products. These companies are time-based competitors.

From Low Wages to Variety Wars

Since 1945, Japanese competitors have shifted their strategic focus at least four times. These early adaptations were straightforward; the shift to time-based competitive advantage is not nearly so obvious. It does, however, represent a logical evolution from the earlier stages.

In the immediate aftermath of World War II, with their economy devastated and the world around them in a shambles, the Japanese concentrated on

George Stalk is vice president of the Boston Consulting Group, where he specializes in time-based competition. With James C. Abbeglen, he wrote Kaisha, the Japanese Corporation: The New Competitors in World Business *(Basic Books, 1985).*

achieving competitive advantage through low labor costs. Since Japan's workers were still productive and the yen was devalued by 98.8% against the dollar, its labor costs were extraordinarily competitive with those of the West's developed economies.

Hungry for foreign exchange, the Japanese government encouraged companies to make the most of their one edge by targeting industries with high labor content: textiles, shipbuilding, and steel–businesses where the low labor rates more than offset low productivity rates. As a result, Japanese companies took market share from their Western competition.

But this situation did not last long. Rising wages, caused by high inflation, combined with fixed exchange rates to erode the advantage. In many industries, manufacturers could not improve their productivity fast enough to offset escalating labor costs. By the early 1960s, for instance, the textile companies–comprising Japan's largest industry–were hard-pressed. Having lost their competitive edge in world markets, they spiraled downward, first losing share, then volume, then profits, and finally position and prestige. While the problem was most severe for the textile business, the rest of Japanese industry suffered as well.

The only course was adaptation: in the early 1960s, the Japanese shifted their strategy, using capital investment to boost work-force productivity. They inaugurated the era of scale-based strategies, achieving high productivity and low costs by building the largest and most capital-intensive facilities that were

> Using focused factories, the Japanese achieved high productivity and low costs.

technologically feasible. Japanese shipbuilders, for example, revolutionized the industry in their effort to raise labor productivity. Adapting fabrication techniques from mass production processes and using automatic and semiautomatic equipment, they constructed vessels in modules. The approach produced two advantages for the Japanese. It drove up their own productivity and simultaneously erected a high capital-investment barrier to others looking to compete in the business.

The search for ways to achieve even higher productivity and lower costs continued, however. And in the mid-1960s, it led top Japanese companies to a new source of competitive advantage–the focused factory. Focused competitors manufactured products either made nowhere else in the world or located in the high-volume segment of a market, often in the heart of their Western competitors' product lines. Focusing of production allowed the Japanese to remain smaller than established broad-line producers, while still achieving higher productivity and lower costs–giving them great competitive power.

Factory costs are very sensitive to the variety of goods a plant produces. Reduction of the product-line

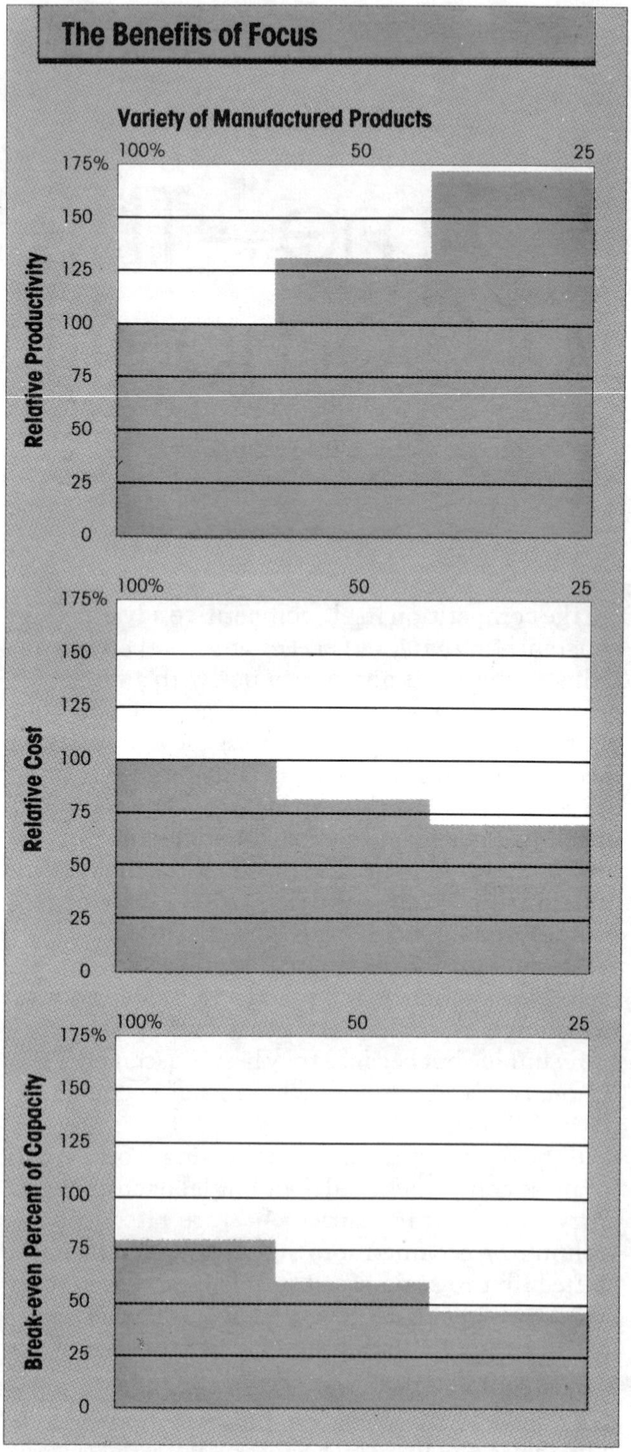

Cutting variety yields higher productivity, lower costs, and reduced break-even points.

variety by half, for example, raises productivity by 30%, cuts costs 17%, and substantially lowers the break-even point. Cutting the product line in half again boosts productivity by 75%, slashes costs 30%, and diminishes the break-even point to below 50%. (See "The Benefits of Focus.")

In industries like bearings, where competition was fierce in the late 1960s, the Japanese fielded product lines with one-half to one-quarter the variety of their Western competitors. Targeting the high-volume segments of the bearing business–bearings for automobile applications was one–the Japanese used the low costs of their highly productive focused factories to undercut the prices of Western competitors.

SKF was one victim. With factories scattered throughout Europe, each geared to a broad product line for the local market, the Swedish company was a big target for the Japanese. SKF reacted by trying to avoid direct competition with the Japanese: it added higher margin products to serve specialized applications. But SKF did not simultaneously drop any low-margin products, thereby complicating its plant operations and adding to production costs. In effect, SKF provided a cost umbrella for the Japanese. As long as they operated beneath it, the Japanese could expand their product line and move into more varied applications.

Avoiding price competition by moving into higher margin products is called margin retreat–a common response to stepped-up competition that eventually leads to corporate suicide. As a company retreats, its costs rise as do its prices, thus "subsidizing" an aggressive competitor's expansion into the vacated position. The retreating company's revenue base stops growing and may eventually shrink to the point where it can no longer support the fixed cost of the operation. Retrenchment, restructuring, and further shrinkage follow in a cycle that leads to inevitable extinction.

SKF avoided this fate by adopting the Japanese strategy. After a review of its factories, the company focused each on those products it was best suited to manufacture. If a product did not fit a particular factory, it was either placed in another, more suitable plant or dropped altogether. This strategy not only halted SKF's retreat but also beat back the Japanese advance.

At the same time, however, leading Japanese manufacturers began to move toward a new source of competitive advantage–the flexible factory. Two developments drove this move. First, as they expanded and penetrated more markets, their narrow product lines began to pinch, limiting their ability to grow. Second, with growth limited, the economics of the focus strategy presented them with an unattractive choice: either reduce variety further or accept the higher costs of broader product lines.

In manufacturing, costs fall into two categories: those that respond to volume or scale and those that are driven by variety. Scale-related costs decline as volume increases, usually falling 15% to 25% per unit each time volume doubles. Variety-related costs, on the other hand, reflect the costs of complexity in manufacturing: setup, materials handling, inventory, and many of the overhead costs of a factory. In most cases, as variety increases, costs increase, usually at a rate of 20% to 35% per unit each time variety doubles.

The sum of the scale- and variety-related costs represents the total cost of manufacturing. With effort, managers can determine the optimum cost point for their factories–the point where the combination of volume and variety yields the lowest total manufacturing cost for a particular plant. When markets are good, companies tend to edge toward increased variety in search of higher volumes, even though this will mean increased costs. When times are tough, companies pare their product lines, cutting variety to reduce costs.

> Flexible manufacturing solved the dilemma; it produced both lower costs and greater variety.

In a flexible factory system, variety-driven costs start lower and increase more slowly as variety grows. Scale costs remain unchanged. Thus the optimum cost point for a flexible factory occurs at a higher volume and with greater variety than for a traditional factory. A gap emerges between the costs of the flexible and the traditional factory–a cost/variety gap that represents the competitive advantage of flexible production. Very simply, a flexible factory enjoys more variety with lower total costs than traditional factories, which are still forced to make the trade-off between scale and variety. (See "The Advantage of Flexible Manufacturing.")

Yanmar Diesel illustrates how this process works. In 1973, with the Japanese economy in recession, Yanmar Diesel was mired in red ink. Worse, there was no promise that once the recession had passed, the existing strategy and program would guarantee real improvement in the company's condition.

As a Toyota supplier, Yanmar was familiar with the automaker's flexible manufacturing system. Moreover, Yanmar was impressed with the automaker's ability to weather the recession without losing

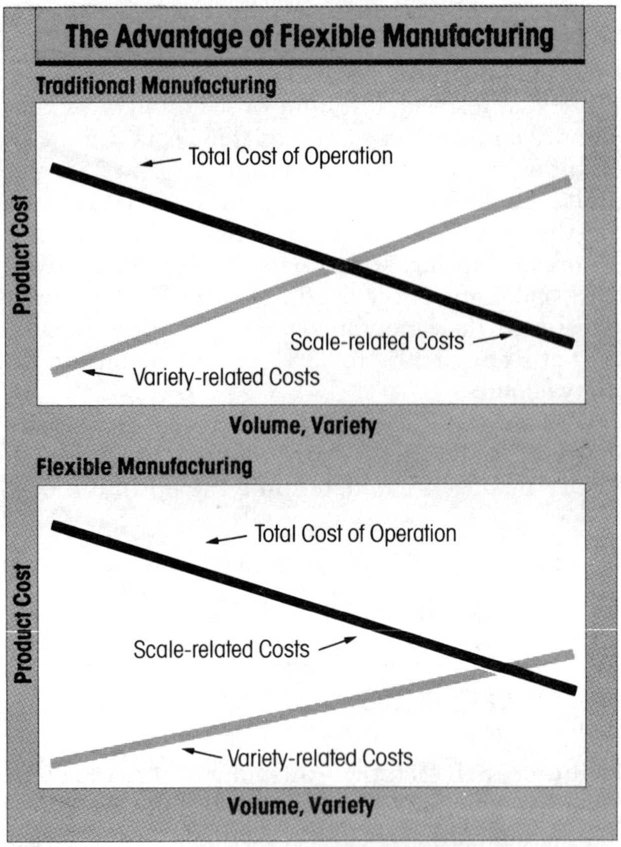

For flexible factories, the optimum cost points occur at a higher volume and with higher variety than for traditional factories.

money. Yanmar decided to install the Toyota procedure in its own two factories. The changeover took less than five years and produced dramatic results: manufacturing costs declined 40% to 60%, depending on the product; factory break-even points dropped 80% to 50%; total manufacturing labor productivity improved by more than 100%.

But it was Yanmar's newfound capability in product variety that signaled the arrival of a unique strategic edge: during the restructuring Yanmar more than quadrupled its product line. With focused factories, Yanmar could have doubled productivity in such a short time only by reducing the breadth of the product line by 75%. The Toyota system made Yanmar's factories more flexible, reducing costs and producing a greater variety of products.

As its inventor, Taiichi Ohno, said, the Toyota production system was "born of the need to make many types of automobiles, in small quantities with the same manufacturing process." With its emphasis on just-in-time production, total quality control, employee decision making on the factory floor, and close supplier relations, the Toyota system gave the many Japanese manufacturers who adopted it in the mid-1970s a distinct competitive advantage.

A comparison of a U.S. company with a Japanese competitor in the manufacture of a particular automotive suspension component illustrates the nature and extent of the Japanese advantage. The U.S. company bases its strategy on scale and focus: it produces 10 million units per year – making it the world's largest producer – and offers only 11 types of finished parts. The Japanese company's strategy, on the other hand, is to exploit flexibility. It is both smaller and less focused: it manufactures only 3.5 million units per year but has 38 types of finished parts.

With one-third the scale and more than three times the product variety, the Japanese company also boasts total labor productivity that is half again that of its American competitor. Moreover, the unit cost of the Japanese manufacturer is less than half that of the U.S. company. But interestingly, the productivity of the Japanese direct laborers is not as high as that of the U.S. workers, a reflection of the difference in scale. The Japanese advantage comes from the productivity of the overhead employees: with one-third the volume and three times the variety, the Japanese company has only one-eighteenth the overhead employees. (See "Flexible Manufacturing's Productivity Edge.")

> Yanmar's flexible factories cut costs and more than quadrupled its product line.

In the late 1970s, Japanese companies exploited flexible manufacturing to the point that a new competitive thrust emerged – the variety war. A classic example of a variety war was the battle that erupted between Honda and Yamaha for supremacy in the motorcycle market, a struggle popularly known in Japanese business circles as the H-Y War. Yamaha ignited the H-Y War in 1981 when it announced the opening of a new factory which would make it the world's largest motorcycle manufacturer, a prestigious position held by Honda. But Honda had been concentrating its corporate resources on the automobile business and away from its motorcycle operation. Now, faced with Yamaha's overt and public challenge, Honda chose to counterattack.

Honda launched its response with the war cry, "Yamaha wo tsubusu!" ("We will crush, squash, slaughter Yamaha!") In the no-holds-barred battle that ensued, Honda cut prices, flooded distribution channels, and boosted advertising expenditures. Most important – and most impressive to consumers – Honda also rapidly increased the rate of change in its product line, using variety to bury Yamaha. At the start of the war, Honda had 60 models of motorcycles. Over the next 18 months, Honda introduced or replaced 113 models, effectively turning over its entire product line twice. Yamaha also began the war with 60 models; it was able to manage only 37 changes in its product line during those 18 months.

Honda's new product introductions devastated Yamaha. First, Honda succeeded in making motorcycle design a matter of fashion, where newness and freshness were important attributes for consumers. Second, Honda raised the technological sophistication of its products, introducing four-valve engines, composites, direct drive, and other new features. Next to a Honda, Yamaha products looked old, unattractive, and out of date. Demand for Yamaha products dried up; in a desperate effort to move them, dealers were forced to price them below cost. But even that didn't work. At the most intense point in the H-Y War, Yamaha had more than 12 months of inventory in its dealers' showrooms. Finally Yamaha surrendered. In a public statement, Yamaha President Eguchi announced, "We want to end the H-Y War. It is our fault. Of course there will be competition in the future but it will be based on a mutual recognition of our respective positions."

Honda didn't go unscathed either. The company's sales and service network was severely disrupted, requiring additional investment before it returned to a stable footing. However, so decisive was its victory that Honda effectively had as much time as it wanted to recover. It had emphatically defended its title as the world's largest motorcycle producer and done so in a way that warned Suzuki and Kawasaki not to challenge that leadership. Variety had won the war.

Time-Based Competitive Advantage

The strength of variety as a competitive weapon raises an interesting question. How could Japanese companies accommodate such rapid rates of change? In Honda's case, there could be only three possible answers. The company did one of the following:

1. Began the development of more than 100 new models 10 to 15 years before the attack.

2. Authorized a sudden, massive spending surge to develop and manufacture products on a crash basis.

3. Used structurally different methods to develop, manufacture, and introduce new products.

In fact, what Honda and other variety-driven competitors pioneered was time-based competitiveness. They managed structural changes that enabled their operations to execute their processes much faster. As a consequence, time became their new source of competitive advantage.

Flexible Manufacturing's Productivity Edge
(Automobile Suspension Component)

	U.S. Competitor	*Japanese Competitor*
Annual Volume	10M	3.5M
Employees		
Direct	107	50
Indirect	135	7
Total	242	57
Annual Units/Employee	43,100	61,400
Types of Finished Parts	11	38
Unit Cost for Comparable Part (index)	$100	$49

(1987 figures)

While time is a basic business performance variable, management seldom monitors its consumption explicitly–almost never with the same precision accorded sales and costs. Yet time is a more critical competitive yardstick than traditional financial measurements.

Today's new-generation companies compete with flexible manufacturing and rapid-response systems, expanding variety and increasing innovation. A company that builds its strategy on this cycle is a more powerful competitor than one with a traditional strategy based on low wages, scale, or focus. These older, cost-based strategies require managers to do whatever is necessary to drive down costs: move production to or source from a low-wage country; build new facilities or consolidate old plants to gain economies of scale; or focus operations down to the most economic subset of activities. These tactics reduce costs but at the expense of responsiveness.

In contrast, strategies based on the cycle of flexible manufacturing, rapid response, expanding variety, and increasing innovation are time based. Factories are close to the customers they serve. Organization structures enable fast responses rather than low costs and control. Companies concentrate on reducing if not eliminating delays and using their response advantages to attract the most profitable customers.

Many–but certainly not all–of today's time-based competitors are Japanese. Some of them are Sony, Matsushita, Sharp, Toyota, Hitachi, NEC, Toshiba, Honda, and Hino; time-based Western companies include Benetton, The Limited, Federal Express, Domino's Pizza, Wilson Art, and McDonald's. For these leading competitors, time has become the overarching measurement of performance. By reducing

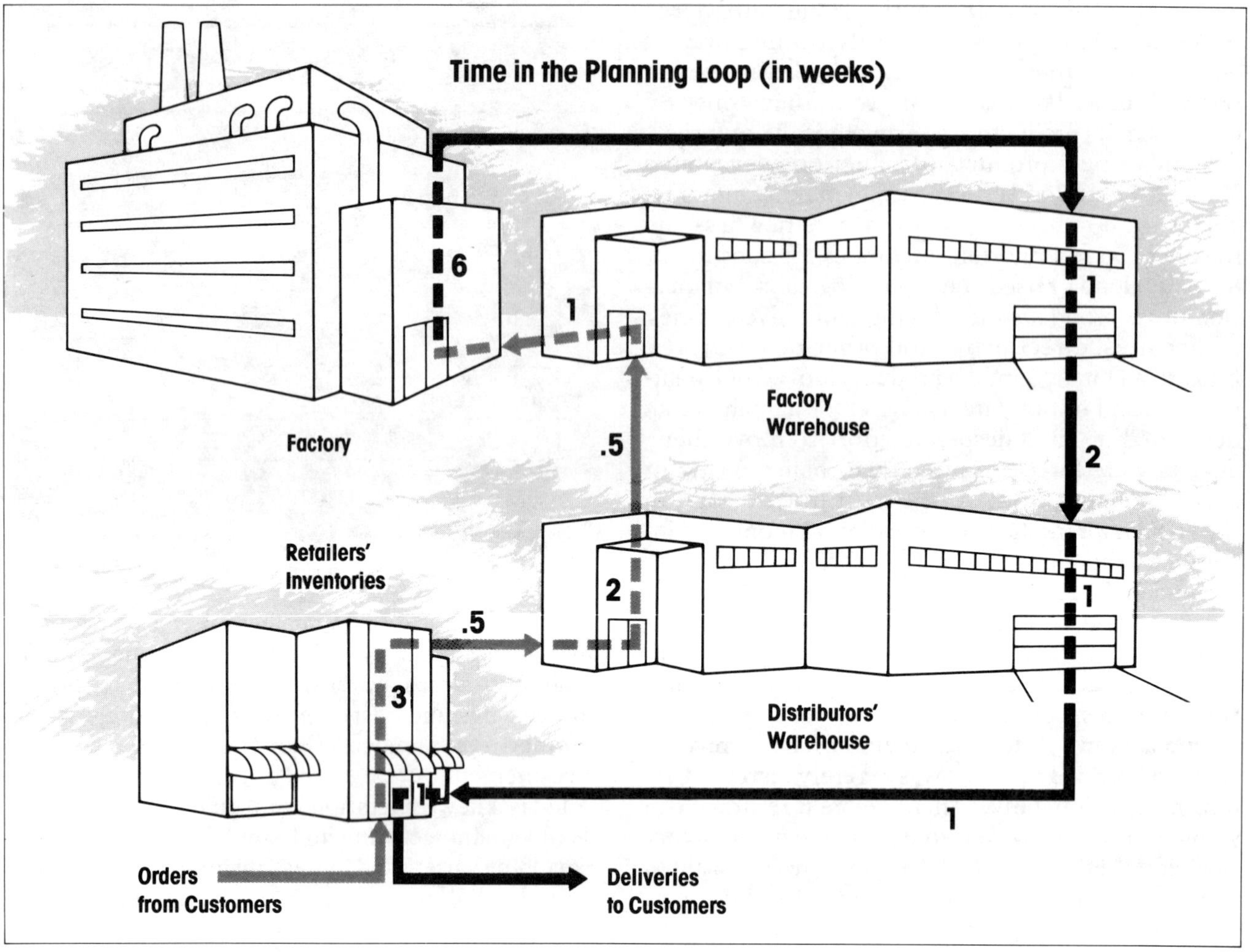

the consumption of time in every aspect of the business, these companies also reduce costs, improve quality, and stay close to their customers.

Breaking the Planning Loop

Companies are systems; time connects all the parts. The most powerful competitors understand this axiom and are breaking the debilitating loop that strangles much of traditional manufacturing planning.

Traditional manufacturing requires long lead times to resolve conflicts between various jobs or activities that require the same resources. The long lead times, in turn, require sales forecasts to guide planning. But sales forecasts are inevitably wrong; by definition they are guesses, however informed. Naturally, as lead times lengthen, the accuracy of sales forecasts declines. With more forecasting errors, inventories balloon and the need for safety stocks at all levels increases. Errors in forecasting also mean more unscheduled jobs that have to be expedited, thereby crowding out scheduled jobs. The need for longer lead times grows even greater and the planning loop expands even more, driving up costs, increasing delays, and creating system inefficiencies.

Managers who find themselves trapped in the planning loop often respond by asking for better forecasts and longer lead times. In other words, they treat the symptoms and worsen the problem. The only way to break the planning loop is to reduce the consumption of time throughout the system; that will, in turn, cut the need for lead time, for estimates, for safety stocks, and all the rest. After all, if a company could ever drive its lead time all the way to zero, it would have to forecast only the next day's sales. While that idea of course is unrealistic, successful time-based competitors in Japan and in the West have kept their lead times from growing and some have even reduced them, thereby diminishing the planning loop's damaging effects.

Thirty years ago, Jay W. Forrester of MIT published a pioneering article in HBR, "Industrial Dynamics: A Major Breakthrough for Decision Makers" (July-August 1958), which established a model of time's impact on an organization's performance. Using "industrial dynamics"–a concept originally developed for shipboard fire control systems–Forrester tracked the effects of time delays and decision rates within a simple business system consisting of a factory, a factory warehouse, a distributors' inventory, and retailers' inventories. The numbers in the illustration "Time in the Planning Loop" are the delays in the flow of information or product, measured in weeks. In this example, the orders accumulate at the retailer for three weeks, are in the mail for half a week, are delayed at the distributor for two weeks, go back into the mail for another half a week, and need eight weeks for processing at the factory and its warehouse. Then the finished product begins its journey back to the retailer. The cycle takes 19 weeks.

> What distorts the system is time: lengthy delays inevitably create an inaccurate view of the market.

The system in this example is very stable–as long as retail demand is stable or as long as forecasts are accurate 19 weeks into the future. But if unexpected changes occur, the system must respond. The chart, also taken from the Forrester article, shows what happens to this system when a simple change takes place: demand goes up 10%, then flattens. Acting on new forecasts and seeking to cut delivery delays, the factory first responds by ramping up production 40%. When management realizes–too late–that it has overshot the mark, it cuts production 30%. Too late again it learns that it has overcorrected. This ramping up and cutting back continue until finally the system stabilizes, more than a year after the initial 10% increase.

What distorts the system so badly is time: the lengthy delay between the event that creates the new demand and the time when the factory finally receives the information. The longer that delay, the more distorted is the view of the market. Those distortions reverberate throughout the system, producing disruption, waste, and inefficiency.

These distortions plague business today. To escape them, companies have a choice: they can produce to forecast or they can reduce the time delays in the flow of information and product through the system. The traditional solution is to produce to forecast. The new approach is to reduce time consumption.

Because time flows throughout the system, focusing on time-based competitive performance results in improvements across the board. Companies generally become time-based competitors by first correcting their manufacturing techniques, then fixing sales and distribution, and finally adjusting their approach to innovation. Ultimately, it becomes the basis for a company's overall strategy.

Time-Based Manufacturing

In general, time-based manufacturing policies and practices differ from those of traditional manufacturers along three key dimensions: length of production runs, organization of process components, and complexity of scheduling procedures.

When it comes to lot size, for instance, traditional factories attempt to maximize production runs while time-based manufacturers try to shorten their production runs as much as possible. In fact, many Japanese companies aim for run lengths of a single unit. The thinking behind this is as simple as it is fundamental to competitive success: reduced run lengths mean more frequent production of the complete mix of products and faster response to customers' demands.

Factory layout also contributes to time-based competitive advantage. Traditional factories are usually organized by process technology centers. For example, metal goods manufacturers organize their factories into shearing, punching, and braking departments; electronic assemblers have stuffing, wave soldering, assembly, testing, and packing departments. Parts move from one process technology center to the next. Each step consumes valuable time: parts sit, waiting to move; then move; then

Unexpected Change Distorts the System

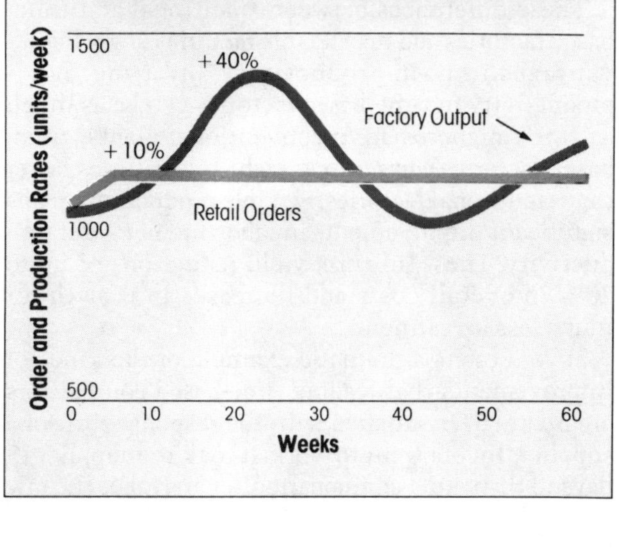

wait to be used in the next step. In a traditional manufacturing system, products usually receive value for only .05% to 2.5% of the time that they are in the factory. The rest of the time products sit waiting for something to happen.

Time-based factories, however, are organized by product. To minimize handling and moving of parts, the manufacturing functions for a component or a product are as close together as possible. Parts move from one activity to the next with little or no delay. Because the production process eliminates the need to pile and repile parts, they flow quickly and efficiently through the factory.

In traditional factories, scheduling is also a source of delay and waste. Most traditional factories use central scheduling that requires sophisticated materials resource planning and shop-floor control systems. Even though these systems are advanced, they still waste time: work orders usually flow to the factory floor on a monthly or weekly basis. In the meantime, parts can sit idle.

In time-based factories, local scheduling enables employees to make more production control decisions on the factory floor, without the time-consuming loop back to management for approval. Moreover, the combination of the product-oriented layout of the factory and local scheduling makes the total production process run more smoothly. Once a part starts through the production run, many of the requirements between manufacturing steps are purely automatic and require no intermediate scheduling.

> Toyota went to work with its too-slow supplier, cutting response time from 15 days to 1.

These differences between traditional and time-based factories add up. Flexible factories enjoy big advantages in both productivity and time: labor productivity in time-based factories can be as much as 200% higher than in conventional plants; time-based factories can respond eight to ten times faster than traditional factories. Flexible production means significant improvements in labor and net-asset productivity. These, in turn, yield reductions of up to 20% in overall costs and increases in growth for much less investment.

Toyota offers a dramatic example of the kinds of improvements that leading time-based competitors are making. Dissatisfied with the response time of a supplier, Toyota went to work. It took the supplier 15 days to turn out a component after arrival of the raw materials at its factory. The first step was to cut lot sizes, reducing response time to 6 days. Next Toyota streamlined the factory layout, reducing the number of inventory holding points. The response time fell to 3 days. Finally Toyota eliminated all work-in-progress inventories at the supplier's plant. New response time: 1 day.

Toyota, of course, is not alone in improving manufacturing response times. Matsushita cut the time needed to make washing machines from 360 hours to just 2; Honda slashed its motorcycle fabricating time by 80%; in North America, companies making motor controllers and electrical components for unit air conditioners have improved their manufacturing response times by 90%.

Time-Based Sales and Distribution

A manufacturer's next challenge is to avoid dissipation of factory performance improvements in other parts of the organization. In Jay Forrester's example of the planning loop, the factory and its warehouse accounted for roughly one-half of the system's time. In actuality today, the factory accounts for one-third to one-half of the total time–often the most "visible" portion of time. But other parts of the system are just as important, if less apparent. For example, in the Forrester system, sales and distribution consume as much or more time than manufacturing.

What Forrester modeled, the Japanese experienced. By the late 1970s, leading Japanese companies were finding that inefficient sales and distribution operations undercut the benefits of their flexible manufacturing systems. Toyota, which at that time was divided into two separate companies, Toyota Motor Manufacturing and Toyota Motor Sales, again makes this point. Toyota Motor Manufacturing could manufacture a car in less than 2 days. But Toyota Motor Sales needed from 15 to 26 days to close the sale, transmit the order to the factory, get the order scheduled, and deliver the car to the customer. By the late 1970s, the cost-conscious, competition-minded engineers at Toyota Manufacturing were angry at their counterparts at Toyota Motor Sales, who were frittering away the advantage gained in the production process. The sales and distribution function was generating 20% to 30% of a car's cost to the customer–more than it cost Toyota to manufacture the car!

Finally, in 1982 Toyota moved decisively to remedy the problem. The company merged Toyota Motor Manufacturing and Toyota Motor Sales. The company announced that it wanted to become "more marketing driven." While Toyota assured the public that the reorganization only returned it to its configuration in the 1950s, within 18 months all the Toyota Motor Sales directors retired. Their jobs were left vacant or filled by executives from Toyota Motor Manufacturing.

The company wasted no time in implementing a plan to cut delays in sales and distribution, reduce costs, and improve customer service. The old system, Toyota found, had handled customer orders in batches. Orders and other crucial information would accumulate at one step of the sales and distribution process before dispatch to the next level, which wasted time and generated extra costs.

To speed the flow of information, Toyota had to reduce the size of the information batches. The solution came from a company-developed computer system that tied its salespeople directly to the factory scheduling operation. This link bypassed several levels of the sales and distribution function and enabled the modified system to operate with very small batches of orders.

Toyota expected this new approach to cut the sales and distribution cycle time in half–from four to six weeks to just two to three weeks across Japan. (For the Tokyo and Osaka regions, which account for roughly two-thirds of Japan's population, the goal was to reduce cycle time to just two days.) But by 1987 Toyota had reduced system responsiveness to eight days, including the time required to make the car. In the Forrester example, this achievement is equivalent to cutting the 19-week cycle to 6 weeks. The results were predictable: shorter sales forecasts, lower costs, happier customers.

Time-Based Innovation

A company that can bring out new products three times faster than its competitors enjoys a huge advantage. Today, in one industry after another, Japanese manufacturers are doing just that to their Western competition:

- In projection television, Japanese producers can develop a new television in one-third the time required by U.S. manufacturers.
- In custom plastic injection molds, Japanese companies can develop the molds in one-third the time of U.S. competitors and at one-third the cost.
- In autos, Japanese companies can develop new products in half the time–and with half as many people–as the U.S. and German competition.

To accomplish their fast-paced innovations, leading Japanese manufacturers have introduced a series of organizational techniques that precisely parallel their approach to flexible manufacturing:

- In manufacturing, the Japanese stress short production runs and small lot sizes. In innovation, they favor smaller increments of improvement in new products, but introduce them more often–versus the Western approach of more significant improvements made less often.
- In the organization of product development work, the Japanese use factory cells that are cross-functional teams. Most Western new product development activity is carried out by functional centers.
- In the scheduling of work, Japanese factories stress local responsibility, just as product development scheduling is decentralized. The Western approach to both requires plodding centralized scheduling, plotting, and tracking.

The effects of this time-based advantage are devastating; quite simply, American companies are losing leadership of technology and innovation–supposedly this country's source of long-term advantage.

> **Unless U.S. companies cut new product cycles by 300%, Japanese manufacturers will easily out-innovate them.**

Unless U.S. companies reduce their new product development and introduction cycles from 36-48 months to 12-18 months, Japanese manufacturers will easily out-innovate and outperform them. Taking the initiative in innovation will require even faster cycle times.

Residential air conditioners illustrate the Japanese ability to introduce more technological innovation in smaller increments–and how in just a few years these improvements add up to remarkably superior products. The Japanese introduce innovations in air conditioners four times faster than their American competitors; in technological sophistication the Japanese products are seven to ten years ahead of U.S. products.

Look at the changes in Mitsubishi Electric's three-horsepower heat pump between 1975 and 1985. From 1975 to 1979, the company did nothing to the product except change the sheet metal work, partly to improve efficiency but mostly to reduce materials costs. In 1979, the technological sophistication of the product was roughly equal to that of the U.S. competition. From this point on, the Japanese first established, and then widened the lead.

In 1980, Mitsubishi introduced its first major improvement: a new product that used integrated circuits to control the air-conditioning cycle. One year later, the company replaced the integrated circuits with microprocessors and added two important innovations to increase consumer demand. The first was "quick connect" freon lines. On the old product (and on the U.S. product), freon lines were made from copper tubing and cut to length, bent, soldered together, purged, and filled with freon–an operation requiring great skill to produce a reliable air conditioner. The Japanese substituted quick-connect freon lines–precharged hoses that simply clicked together. The second innovation was simplified wiring. On the old product (and still today on the U.S. product) the unit had six color-coded wires to connect. The advent of microprocessors made possible a two-wire connection with neutral polarity.

These two changes did not improve the energy-efficiency ratio of the product; nor were they intended to. Rather, the point was to fabricate a unit that would be simpler to install and more reliable, thereby broadening distribution and increasing demand. Because of these innovations, white-goods outlets could sell the new product, and local contractors could easily install it.

In 1982, Mitsubishi introduced a new version of the air conditioner featuring technological advances related to performance. A high-efficiency rotary compressor replaced the outdated reciprocating compressor. The condensing unit had louvered fins and inner fin tubes for better heat transfer. Because the balance of the system changed, all the electronics had to change. As a result, the energy-efficiency ratio improved markedly.

In 1983, Mitsubishi added sensors to the unit and more computing power, expanding the electronic control of the cycle and again improving the energy-efficiency ratio.

Mitsubishi's time-based approach left U.S. air-conditioner companies ten years behind.

In 1984, Mitsubishi came out with another version of the product, this time with an inverter that made possible an even higher energy-efficiency ratio. The inverter, which requires additional electronics for the unit, allows unparalleled control over the speed of the electric motor, dramatically boosting the appliance's efficiency.

Using time-based innovation, Mitsubishi transformed its air conditioner. The changes came incrementally and steadily. Overall they gave Mitsubishi–and other Japanese companies on the same track–the position of technological leadership in the global residential air-conditioning industry.

In 1985, a U.S. air-conditioner manufacturer was just debating whether to use integrated circuits in its residential heat pump. In view of its four- to five-year product development cycle, it could not have introduced the innovation until 1989 or 1990–putting the American company ten years behind the Japanese. Faced with this situation, the U.S. air-conditioner company followed the example of many U.S. manufacturers that have lost the lead in technology and innovation: it decided to source its air conditioners and components from its Japanese competition.

Time-Based Strategy

The possibility of establishing a response time advantage opens new avenues for constructing winning competitive strategies. At most companies, strategic choices are limited to three options:

1. Seeking coexistence with competitors. This choice is seldom stable, since competitors refuse to cooperate and stay put.

2. Retreating in the face of competitors. Many companies choose this course; the business press fills its pages with accounts of companies retreating by consolidating plants, focusing their operations, outsourcing, divesting businesses, pulling out of markets, or moving upscale.

3. Attacking, either directly or indirectly. The direct attack involves the classic confrontation–cut price and add capacity, creating head-on competition. Indirect attack requires surprise. Competitors either do not understand the strategies being used against them or they do understand but cannot respond–sometimes because of the speed of the attack, sometimes because of their inability to mount a response.

Of the three options, only an attack creates the opportunity for real growth. Direct attack demands superior resources; it is always expensive and potentially disastrous. Indirect attack promises the most gain for the least cost. Time-based strategy offers a powerful new approach for successful indirect attacks against larger, established competitors.

Consider the remarkable example of Atlas Door, a ten-year-old U.S. company. It has grown at an average annual rate of 15% in an industry with an overall annual growth rate of less than 5%. In recent years, its pretax earnings were 20% of sales, about five times the industry average. Atlas is debt free. In its tenth year the company achieved the number one competitive position in its industry.

The company's product: industrial doors. It is a product with almost infinite variety, involving limitless choices of width and height and material. Because of the importance of variety, inventory is almost useless in meeting customer orders; most doors can be manufactured only after the order has been placed.

Historically, the industry had needed almost four months to respond to an order for a door that was out of stock or customized. Atlas's strategic advantage was time: it could respond in weeks to any order. It had structured its order-entry, engineering, manufacturing, and logistics systems to move information and products quickly and reliably.

First, Atlas built just-in-time factories. These are fairly simple in concept. They require extra tooling and machinery to reduce changeover times and a fabrication process organized by product and scheduled to start and complete all of the parts at the same time. But even the performance of the factory–critical to the company's overall responsiveness–still only accounted for 2½ weeks of the completed product delivery cycle.

Second, Atlas compressed time at the front end of the system, where the order first entered and was processed. Traditionally, when customers, distributors, or salespeople called a door manufacturer with a request for price and delivery, they would have to wait more than one week for a response. If the desired door was not in stock, not in the schedule, or not engineered, the supplier's organization would waste even more time, pushing the search for an answer around the system.

> The industry needed almost four months to respond. Atlas could fill any order in just weeks.

Recognizing the opportunity to cut deeply into the time expenditure in this part of the system, Atlas first streamlined, then automated its entire order-entry, engineering, pricing, and scheduling processes. Today Atlas can price and schedule 95% of its incoming orders while the callers are still on the telephone. It can quickly engineer new special orders because it has preserved on computer the design and production data of all previous special orders–which drastically reduces the amount of re-engineering necessary.

Third, Atlas tightly controlled logistics so that it always shipped only fully complete orders to construction sites. Orders require many components. Gathering all of them at the factory and making sure that they are with the correct order can be a time-consuming task. It is even more time-consuming, however, to get the correct parts to the job site *after* they have missed the initial shipment. Atlas developed a system to track the parts in production and the purchased parts for each order, ensuring arrival of all necessary parts at the shipping dock in time–a just-in-time logistics operation.

When Atlas started operations, distributors were uninterested in its product. The established distributors already carried the door line of a larger competitor; they saw no reason to switch suppliers except, perhaps, for a major price concession. But as a start-up, Atlas was too small to compete on price alone. Instead, it positioned itself as the door supplier of last resort, the company people came to if the established supplier could not deliver or missed a key date.

Of course, with industry lead times of almost four months, some calls inevitably came to Atlas. And when it did get a call, Atlas commanded a higher price because of its faster delivery. Atlas not only got a higher price but its time-based processes also yielded lower costs: it thus enjoyed the best of both worlds.

In ten short years, the company replaced the leading door suppliers in 80% of the distributors in the country. With its strategic advantage the company could be selective, becoming the house supplier for only the strongest distributors.

In the wake of this indirect attack, the established competitors have not responded effectively. The conventional view is that Atlas is a "garage shop operator" that cannot sustain its growth: competitors expect the company's performance to degrade to the industry average as it grows larger. But this response–or nonresponse–only reflects a fundamental lack of understanding of time as the source of competitive advantage. The extra delay in responding only adds to the insurmountable lead the indirect time-based attack has created. While the traditional companies track costs and size, the new competitor derives advantage from time, staying on the cutting edge, leaving its rivals behind.

Reprint 88410

Harvard Business Review
July-August 1989

The Service Factory

by Richard B. Chase and David A. Garvin

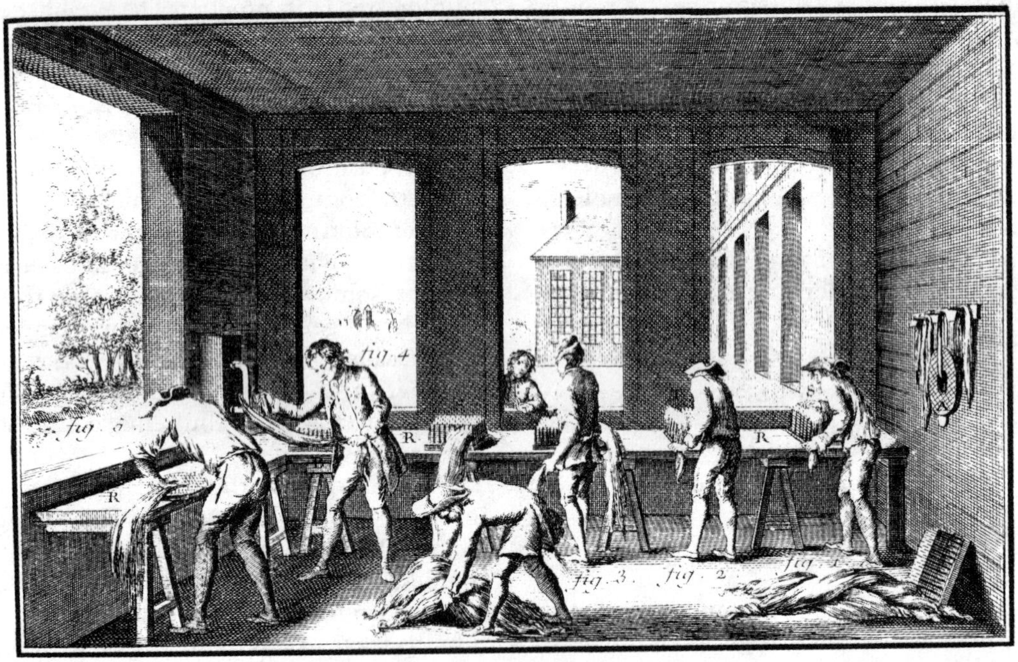

The factory of the future is *not* a place where computers, robots, and flexible machines do the drudge work. That is the factory of the present, which, with money and brains, any manufacturing business can build. Of course, any competitor can build one too – which is why it is becoming harder and harder to compete on manufacturing excellence alone. Lower costs, higher quality, and greater product variety are like table stakes in poker – the price

Richard B. Chase is professor and director of the Center for Operations Management Education and Research at the School of Business Administration at the University of Southern California. David A. Garvin is professor of business administration at the Harvard Business School and faculty chairman of the School's executive program, Manufacturing in Corporate Strategy.

that companies pay to enter the game. Most products can be quickly and easily imitated; and the most automated design and production processes cannot decisively beat the second most automated. Who wins and who loses will be determined by how companies play, not simply by the product or process technologies that qualify them to compete.

The manufacturers that thrive into the next generation, then, will compete by bundling services with products, anticipating and responding to a truly comprehensive range of customer needs. Moreover, they will make the factory itself the hub of their efforts to get and hold customers – activities that now are located in separate, often distant, parts of the organization. Production workers and factory managers will be able to forge and sustain new relationships with customers because they will be in direct and continuing contact with them. Manufacturing, in short, will become the cortex of the business. Today's flexible factories will become tomorrow's service factories.

About 200 years ago, when horse-drawn carriages were made largely by craftsmen, the most successful carriage maker was invariably the most accommodating. Though he prided himself on being a technician – a manufacturer – his success depended heavily on his willingness and ability to talk with customers at key points: before the sale, so he could get a clear idea of what the client needed and what features would satisfy him; during the manufacturing process, so he could incorporate any necessary changes in the product; and after delivery, so he could learn what features had worked (and what hadn't) and what the client needed for maintenance, repair, and replacement.

Even the most flexible factory does not yet capitalize on the full range of services manufacturing could provide.

Mass production overtook customized craftsmanship because customers came to value standardized goods over higher priced, personalized goods. As a result, work grew increasingly compartmentalized through the division of labor. Craftsmanship (that is, manufacturing) became separated from downstream activities, like sales and postpurchase service, as well as from upstream activities, like new-product development and design. Gradually, manufacturing received more and more of its information and instructions through filters – divisions and departments that were separated, functionally and physically, from the production site. Not surprisingly, manufacturing managers complained that those who defined their work rarely understood it or cared enough about its details, problems, or technical possibilities.

For decades, companies muddled through. In recent years, as Japanese competition put pressure on manufacturing businesses everywhere, manufacturers have worked mightily and successfully to educate workers and break down some of the barriers between their upstream activities and the work of the factory. They have encouraged interfunctional communication between product designers and manufacturing engineers and between R&D and quality managers on the factory floor.

These imaginative efforts to accelerate product innovation and improve manufacturing performance were necessary and important. But they are no longer adequate. Today downstream activities have to be joined to the tasks of the factory too. Increasingly, factory personnel have the means to support the sales force, service technicians, and consumers. This support should, and will, be used. Competition is shifting away from how companies build their products to how well they serve customers before and after they build them.

Some of America's best-run companies–Hewlett-Packard, Allen-Bradley, Caterpillar, Frito-Lay–already operate factories whose activities reflect the new role of service in manufacturing competition. None of their facilities is a complete service factory. We are still many years from that. But in the range of upstream and downstream activities these factories perform, and in the degree of interaction between production workers and customers, they point the way to the future.

Service for a manufacturing company inescapably revolves around its products–their design, features, durability, repairability, distribution, and ease of installation and use. Even the most traditional factories of yesterday proffered service of a kind, but their conception of service was narrow. To old-guard factory managers, service was little more than a commitment to meeting due dates. Logistics and distribution urged the factory to complete orders in a timely fashion, to give advance notice of delivery problems, and to package materials for ease of shipment and damage control. Customers were simply numbers on a production schedule.

The barriers between factories and upstream product designers have fallen. Break the barriers to downstream customers too.

Managers in flexible factories have broadened the concept of service to include both a commitment to product variety and a capacity to see the customer in specialized terms. The key to flexibility has been the ability to produce small lot sizes efficiently and to change over quickly from one product to another. Yet even the most flexible factories are not capitalizing on the full range of services they could provide. As a result, they are bypassing competitive opportunities.

Factories generate information and skills that are critical for product design. Among them are accurate and timely feedback on the manufacturability of new designs, the ability to construct prototypes quickly, and the capability of introducing engineering change orders smoothly. But factories are also a resource for helping customers with installation, maintenance, and troubleshooting. The people who make products are often more knowledgeable about their performance, variability, and repair than the people in field service.

Such product experts can also contribute decisively to sales and marketing efforts. Not only do factories generate quality information that helps sales, but producers of very sophisticated products–chip makers, for instance–prove the quality of their products by showing customers that their procedures are virtually infallible. Factories can serve as real-time demonstrators of the technology and systems the company sells. They can also help make salespeople more responsive by supplying detailed cost and production data for competitive bidding and cross-selling and by making an ever-increasing range of products. Once the products have been sold, the factory can also participate in after-sales support by providing accurate and easily accessible information about the status of orders and by replenishing critical parts and products without delay.

Building a full-fledged service factory–one that offers all these services–in one fell swoop would be difficult. The way to reach that goal is to start by identifying a focused set of customer needs and then try to fill them with all deliberate speed, beginning with those services the factory is particularly good at. This means coalescing around a single service model that is so explicit and transparent that the whole organization can easily understand what needs to be done and why.

Here we propose four alternatives; there are undoubtedly more.

The Laboratory. The factory's competence in testing new products and processes and in ensuring manufacturability and quality equips it also to serve as a lab. The archetype is the pilot plant. Factory managers make crucial information available to the rest of the business: data on product performance and architecture to R&D; process parameters to designers; capacity restrictions to sales and marketing. To do so, they need processes that are heavily instrumented, equipment that is capable of fine adjustments, operators who are technically skilled, and documentation that is careful and precise.

Who better than factory personnel to help customers learn the uses and maintenance of products?

Allegheny Ludlum, a large manufacturer of specialty steel, has developed a formal system of process experiments, often high-risk experiments governed by special procedures to encourage innovation. High-risk experiments are those that push the frontiers of steel making; more traditional experiments aim to modify or improve existing operations that remain imperfectly understood.

For example, in the mid-1980s, Allegheny Ludlum began a series of experiments to develop a metallic replacement for the ceramic substrate in automotive catalytic converters. The experiments involved a new family of alloys, which are difficult and time-consuming to manufacture. To begin the experiments, R&D or manufacturing people first obtained sign-offs from the senior vice president of the technical division and the vice presidents of technical services, production, and sales; then, operating managers ran the experiments freely, knowing that the results would not appear in their budgets or detract from their margins.

The company supports such experimentation with an extremely detailed cost accounting system – 100,000 standards, tracking costs for every department, machine, process, grade, width, and gauge of steel. As a result, every experiment yields useful cost and productivity information. The system also tracks defect rates and other quality data. Experimentation has produced impressive results: between 1984 and 1988, the productivity of Allegheny's melt shop and continuous caster improved 7% to 8% per year.

The CEO of Chaparral Steel, a highly successful mini-mill, is explicit about this approach. He is trying, he says, to get his company's research and development done right on the factory floor. First-line supervisors enjoy regular sabbaticals, during which they visit other steel mills, explore new technologies, and visit customers. They bring fresh, often experimental ideas back to the factory and take the lead in introducing them to the shop floor.

Laboratories do more, however, than perform experiments. They also supply supporting data based on rigorous tests – like the clinical labs that support physicians. Such information can also alleviate customers' concerns about product failures and flawed performance. Because customers are less patient and more insistent than ever, the factory that's close to them and does its own product experimentation can respond to their concerns quickly and directly by providing hard data with which they can make judgments about product quality, flexibility, and the like.

At Hewlett-Packard's Fort Collins Systems Division, which makes computers and technical workstations, the quality department has played a lead role in generating information crucial to customers. At first, quality managers developed customer surveys and collected field data on competitors' products for their own use. Then they discovered that marketing viewed the initiative as an encroachment. So they suggested forming a new partnership drawing on the factory's special expertise. The quality department embraced the role of a service department for marketing, working with members of the technical marketing staff to support field sales and service personnel. According to Henry Kohoutek, the quality manager: "The focus for selecting opportunities to contribute was shifted to the marketing staff, who knew well where help was needed and would be appreciated."

Over time, new kinds of information were generated, packaged, and presented in novel formats: product-quality data sheets; test conditions and test-result data summarized in easily understandable ways that impressed customers; videotapes documenting actual product tests and field performance. The quality department also cultivated direct contacts with sales and service personnel through training sessions, presentations, and guided tours through its facilities. Marketing now covers nearly all the associated costs from its own budget.

The Consultant. Hewlett-Packard's experience is closely related to another factory service–problem solving out in the field. Many companies sell technical expertise along with the product. Most use so-called "systems engineers" or technical service people. Yet factory workers can constitute a pool of talent that can perform some of these tasks. The more factory people work with design, marketing, and quality, and the greater the sophistication of the machines they run, the more they come to think like product and process engineers. In the words of one product designer we know, they become "experts on what the product is trying to be." Factory people can teach customers quality control techniques. They could also be made a part of a customer's design team, consulting on manufacturability, or could help customers solve production difficulties.

If customers have questions about Tektronix's oscilloscopes, they can call an "800" number that rings on the factory floor.

Raritan River Steel Company, a metal-rod fabricator, sends factory workers out with salespeople to troubleshoot customers' quality problems; after returning to the factory, they develop solutions with coworkers on the shop floor. In one case, a customer's problem was traceable to a subtle manufacturing flaw–minute scratches on the surface of a rod–that only a worker's trained eye could identify.

Tektronix, a manufacturer of electronic equipment, has pioneered direct communication between customers and shop-floor employees. Into the shipping carton of every oscilloscope it sells, the company inserts a postcard listing the names of the workers who built the scope along with an "800" number to a phone on the shop floor. Every day the factory gets several calls from customers; the six people working in the repair area who answer them have all received telephone training.

Customers call for various reasons: questions about the use of their oscilloscopes, complaints about quality performance, requests for information about other Tektronix products, and so on. Workers and managers meet daily to discuss these calls; if necessary, further conversations with the customer follow up the meetings. In some cases, workers will call customers six months after delivery to find out how well their products are performing.

Of course, to be useful consultants, factory workers must know more about their products and markets than they can absorb by daily osmosis. They should get training in basic quality assurance techniques and education in key technical concepts and terminology. They must also be at ease dealing with customers and even perhaps in making formal presentations.

The Showroom. Another, somewhat unexpected way a factory can be used to advantage, especially in high-tech businesses, is as a showroom. The factory can serve as a working demonstration of the systems, processes, and products it manufactures. It can also dramatize the company's manufacturing superiority and, by implication, its superior quality or reliability.

Allen-Bradley, a manufacturer of industrial automation controls, uses its computer-integrated manufacturing operation in Milwaukee in just these ways. The facility is a 150 feet by 300 feet "factory within a factory." Within 24 hours of order placement, it can produce 1,025 different electronic contactors and relays in lot sizes down to one and with zero defects.

Even before this facility was constructed, the company's sales group had developed a "productivity pyramid" to show the linkages among the various levels of factory control (computer controls at higher and lower levels–plan level, center level, and so forth) and had seized on the idea of using the company's own operations to prove the system's plausibility. The Milwaukee plant also shows off Allen-Bradley's software products, and it models what would otherwise be an abstract and elusive systems architecture.

Factory responsiveness is the linchpin of after-sales support.

The factory can also reinforce customer perceptions of product quality. Copeland Corporation, a manufacturer of air-conditioning compressors, introduced a new model in the 1970s and built a state-of-the-art facility equipped with programmable controllers and other advanced automation. The corporation then arranged customer tours to demonstrate the plant's precision manufacturing and exhaustive test procedures and to underline the product's reliability. Copeland seized 25% of the market in two years.

For the factory to be a successful showroom, three things are needed. Marketing and manufacturing personnel have to work closely together to both understand customers' expectations and fulfill them. Manufacturing managers and shop-floor employees have to be well trained in communication and presentation. And the factory's layout must allow for stopping points and audiovisual aids to highlight processes with the greatest sales potential.

At Frito-Lay's Vancouver, Washington plant, for example, entrances and exits have been enlarged to accommodate large groups of visitors, and microphones and loudspeakers have been located at key observation points. Floors have been reconditioned for safety, and public affairs people have developed three different tours, one tailored to wholesalers, one to retailers, and, of course, one to snackers. One of these tours was critical to landing a $2 million account.

The Dispatcher. Finally, the factory can be the linchpin of after-sales support. Service backup is of obvious importance in emerging industries, where the technology is new and products lack a track record. But factories can also be organized to help companies differentiate products in mature businesses. They can provide customers with quick replacements for defective or worn parts and ensure speedy replenishment of stocks to help customers avoid downtime and stock-outs. An unavailable part can idle an expensive machine – indeed, an entire factory.

Effectiveness as a dispatcher requires an unprecedented level of flexibility. The factory must be able to anticipate demand surges and respond immediately to customers' crises.

What connects factories and customers is strategy, not technology.

For decades, Caterpillar has been extraordinarily skilled at parts control, warehousing, and logistics; it promises to have repair parts for its tractors and earth-moving equipment available anywhere in the world within 48 hours. This responsiveness is emphasized more than ever by Caterpillar's sales force. Today Caterpillar often advises customers in the area of logistical support and related services and has an active consulting practice.

In clothing, personal care, and small appliance businesses, which are often subject to fads and shifts in fashion, quick responsiveness to the retail trade is crucial. The Limited, a clothing retailer, electronically links its hundreds of stores to a single, centralized computer at its Columbus, Ohio headquarters and then to its Hong Kong textile mills through a real-time system that allows the mills to begin reproducing hot-selling items at the end of the first week of sales.

Such responsiveness depends on ever-tighter customer linkages, which can be achieved in several ways. In the most innovative approach, sometimes called interorganizational systems, factories supply customers with computer terminals that are linked directly to the factory's order-entry and production-control system. Inland Steel uses interorganizational systems, as does Levi Strauss. Several leading food companies are exploring the possibility of linking their order-entry systems to the computers that compile data from supermarket scanners so that they can bypass the purchasing departments entirely.

Companies can also tighten customer linkages more conventionally, without sophisticated information systems. Usually, this requires interfunctional teams consisting of representatives from sales, logistics, and manufacturing. At Frito-Lay's Vancouver plant, for instance, logistics and sales managers and plant personnel schedule regular (weekly, monthly, quarterly) contacts, in person and by telephone. Joint training events are common. Teams work together to reduce late deliveries, eliminate stale product, resolve breakage problems, and meet special promotion needs. The result has been rapid sales growth and, in 1988, the highest combined service and quality rating of any of Frito-Lay's full-mix plants.

Each of these models – laboratory, consultant, showroom, and dispatcher – shows a distinctive approach to factory service. They also show how services overlap. A factory cannot become a good dispatcher without excellent customer information; it cannot become a good consultant without detailed product and process knowledge. Factories become service factories when their managers and workers understand customers' needs as deeply and directly as they know their own products.

Not all factory managers welcome the changes a service factory requires. Most companies used to seal off their core production technology and discouraged interaction with groups outside the factory. Manufacturing managers sought to maximize efficiency and protect the line from outside disturbances; they buffered themselves by storing inventories in locations that were set off from the rest of the organization and from customers.

The managers of service factories, in contrast, have to work in an open system. They need direct and accessible connections to design, to marketing, and to strategic planning, as well as to customers. Computers and telecommunications can help here, to be sure. Computerized ordering systems, expert systems to manage complex sales, computerized logs for after-sales support, computerized catalogs for replacement parts, 24-hour answering machines to take customer complaints, inexpensive fax machines – all of these speed up communication and break down functional barriers.

But what drives the open systems approach is strategy, not technology – the fact that management sees the competitive advantages of working more collaboratively. At Frito-Lay's Vancouver plant, the best example of an open system we've come across, the technology is nothing more than a telephone.

Tomorrow's leading manufacturing companies will be the ones whose managers unleash the service potential of their factories. Competition demands it. Technology makes it possible. Indeed, the competitive forces driving companies to differentiate products with new and imaginative services are simultaneously empowering factory organizations to deliver them.

Think again about the division of labor, how the factory organization has evolved – and why. Victorian capitalists (and, later, the "scientific management" movement led by Frederick Winslow Taylor) divided human labor into ever more simplified and automatic tasks. The goal was improved productivity: if every worker did a small part of the job, unskilled workers could build products more quickly and efficiently (if less pleasurably) than craftsmen. The point was to compete on price.

The forces that are driving manufacturing to compete on services also empower the factory to deliver them.

The Japanese raised this principle to a higher, more sophisticated standard. Taking their cues from Taylor and his disciples, Japanese managers have analyzed production processes with astonishing precision – movements in microseconds, tolerances in fractions of microns. Originally, the idea was simply to make people work more efficiently. But this attention to detail also laid the groundwork for just-in-time and computerized manufacturing. Japanese factories were organized to deliver low-cost products fast without sacrificing superior quality or flexibility. (You could always buy a luxurious, indestructible car in just the right color. But not off the lot, and not for three months' wages.)

To stay in the game, U.S. manufacturers have had to reform their processes and practices. Many have succeeded. Today Harley-Davidson stands up to Honda because of its embrace of quality improvements and just-in-time production methods, when a few years ago the company was on the brink of financial disaster. Every business now has to master the science of manufacturing – the analysis, subdivision, and control of tightly defined conversion tasks. Otherwise, its factories will remain hopelessly unfit for world competition and for the computer programs that might run and monitor the line.

Moreover, U.S. industry now has to select, train, and retrain workers who use and compete with smart (and not so smart) machines. Today direct labor averages less than 15% of the cost of most manufactured goods; in five years that number is likely to seem as extravagant as 3% defect rates recently did.

As factories employ fewer people, the ones who remain must contribute as no computer or robot can.

As factories employ fewer and fewer people, those who remain need to know how to deal with complex machines, software interfaces, and design problems, and how to track quality and appreciate customers. Increasingly, as Peter Drucker and others have argued, the factory requires knowledge workers who will add value by thinking more like general managers, by contributing, as no computer can, by seeing the production system whole and suggesting fresh ways to enhance products. The result is that factory workers and managers will know more about selected aspects of their products' performance and potential than anyone else in the organization.

Drucker has noted that the imperatives of information-based competition are breaking down barriers within businesses and making functional divisions obsolete. He might have added that divisions among upstream and downstream activities are evaporating too. Marketing, research and development, design, manufacturing, sales, after-sales service–all are folding into one another. The service tasks of business can no longer be separated neatly and sequentially from the work of the factory. If factory personnel are indispensible to the interfunctional teams that generate excellent designs–as they are–how much more essential they must be to a business that competes on service.

Reprint 89402

Manufacturers need kanban for daily operations and MRP II for long-term planning.

Getting Control of Just-in-Time

by Uday Karmarkar

Like all good revolutions, just-in-time manufacturing is producing revolutionaries who don't know when to stop. It is also producing overreactions from people determined to make them stop. Consider the curiously vexed debate about how to get materials to, and work in process through, the shop floor.

Pick up virtually any manufacturing magazine these days and there will be some articles and pages of advertisements by consultants extolling the virtues of JIT over such computer-driven control systems as materials requirements planning (MRP) or materials resource planning (MRP II) – as if JIT principles were opposed to MRP and to the use of computers. One recent ad put the choice this starkly: "JIT vs. MRP II – JIT is the key to your survival!"

MRP, proponents of JIT explain, is merely a "push" technique. An MRP II program promises manufacturing managers more precision than it can deliver, requires unnecessary information, and demands more formal discipline than the shop floor needs. In contrast, JIT people seem especially drawn to such computerless, "pull" techniques as kanban, the system used extensively in Japan's auto and electronics industries. For JIT, presumably, human pull is good, computer push is bad.

What must be particularly confusing to manufacturing managers who get wind of this debate is that kanban systems are used most successfully by the same Japanese and American corporations that are famous for spearheading the use of advanced computer automation – Toyota or Hewlett-Packard, for example. In crucial respects, MRP II aims to be a JIT system, while kanban cannot. Worst of all, doesn't kanban look suspiciously like the old order point, order quantity (OP, OQ) system that MRP once discredited and replaced?

Uday Karmarkar is the Xerox Professor of Operations Management and director of the Center for Manufacturing and Operations Management at the William E. Simon Graduate School of Business Administration of the University of Rochester.

DRAWINGS BY MICHAEL WITTE

This debate needs clarity, and then it needs to end. The idealized conception of the shop floor one gets from some extreme JIT advocates – a line, inherently flexible, inventoryless, even computerless, replenished by infinitely responsive suppliers – may actually prevent manufacturing managers from using the tools they need to run their operations. JIT principles should certainly not preempt the use of MRP II. Indeed, most advanced manufacturing companies find that they require a hybrid system of shop floor control systems – tailored systems, including innovative pull systems like kanban, as well as time-tested, computer-driven push systems like MRP II.

At the same time, shop floor managers should know just when MRP II is an unnecessary burden and when kanban can't work – when push comes to shove and pull comes to tug. All managers can learn interesting strategic lessons from their choices. The question of how to manage inventory cuts quickly to the basics of manufacturing in an age of intense global competition: How much automation is enough? How should the factory respond to customers? How much can you load on workers? How do you deal with orders? What is waste? The shop floor is still a microcosm of the whole business.

Pulling and Pushing at JIT

The basic difference between pull and push is that a pull system initiates production as a reaction to present demand, while push initiates production in anticipation of future demand. Thus a fast food restaurant like McDonald's runs on a pull system, while a catering service operates a push system.

At McDonald's, the customer orders a hamburger, the server gets one from the rack, the hamburger maker keeps an eye on the rack and makes new burgers when the number gets too low. The manager orders more ground beef when the maker's inventory gets too low. In effect, the customer's purchase triggers the pull of materials through the system; the customer initiates a chain of demand.

In a push system, the caterer estimates how many steaks or lobsters are likely to be ordered in any given week. He reckons roughly how long it takes to broil a steak or serve a party of four; he figures out how many meals he can accommodate and commits to buying what he needs in advance. He can account for the special event that he knows is scheduled to take place in midweek. In other words, the caterer gets a picture of production in his mind and pushes materials to where he expects them to be needed.

JIT extremists see human "pull" as good and computer "push" as bad.

What Is JIT Anyway? When you drop everything to take care of a walk-in client, you are reacting, implicitly, to a pull system. When you plan for a meeting, you are in the push mode. What does either method have to do with JIT?

Nothing directly. Think of JIT as a statement of objectives. It underscores the importance of lead-time management in all aspects of manufacturing. It asserts that incremental reductions in lead times are crucial indices of manufacturing improvement. JIT presumes that to achieve such reductions the system should deliver to every operator, in any conversion process, whatever he or she needs just when it is needed. It saves the money tied up in downstream inventories, protecting against long lead times. Shorter lead times mean improved responsiveness and flexibility.

JIT promises to preempt the delays and confusion associated with the stack-up of materials. Correspondingly, it saves the money that would other-

wise go into indirect labor for storing and moving work-in-process (WIP) inventory and storing and handling buffer stock.

An analogy to commuter traffic may be useful. Say a thousand cars have to get through the Lincoln Tunnel every ten minutes. Wouldn't it be ideal for the economy (and sanity) of New Yorkers if each driver left home at precisely the right time to fall into the line at the entrance, so each car entered, one after another, like cars in a train? Would New Yorkers need six-lane divided highways leading to the tunnel if they could enforce this JIT ideal?

To be sure, there are a number of established materials control techniques – pull techniques – associated with JIT, particularly as Japanese companies have realized it. There are synchronized deliveries from proximate suppliers, who can also deliver more or less as production needs fluctuate. There are floor layouts and kanban systems in which materials are constrained to flow consecutively along predictable paths and at a pace determined, in effect, by the last operator in the chain.

These pull techniques, excellent as they are, should not, however, be confused with JIT's governing principles. JIT aims to manage lead times and eliminate waste. There is nothing inherent in push systems that makes them incompatible with JIT. On the contrary. The goal is to get drivers to leave home at just the right time to take their places at the tunnel entrance. JIT advocates shouldn't care whether the signal to leave comes as a computerized call or as a wave from the next-door neighbor.

The Limits of Pull. There is no denying that pull systems are very effective in disciplining production to meet demand just-in-time and, in certain environments, are more effective than push systems. But are pull systems inherently JIT? No pull system, in fact, can constrain workers consciously to produce just-in-time for some future event – because pull systems do not recognize future events. The inventory level triggers production; the pull system aims to fill up depleted inventory – whether it's Big Macs or machined parts.

Pull systems are fine if your McDonald's franchise is downtown with a steady daily stream of customers. But if you're next to a football stadium, how can a pull system alone prepare you for the day of a game? Similarly, it's easy to see how implausible a pull system would be in solving the Lincoln Tunnel's JIT problem. Wouldn't it be more sensible to stagger the cars according to some kind of push system – say, "blue cars leave at 7:45" – than to expect all commuters acting individually to leave for work in perfect synchrony?

There is a paradox here. JIT advocates admire pull systems and look askance at computer-governed push systems like MRP. Yet the latter inherently aims to be a JIT system, while pull systems do not really recognize the future events you are supposed to be just-in-time for. This confusion has a history.

Forty years ago, the most common control system was OP, OQ, a pull system that seeks to exploit the presumed efficiencies of batch manufacturing. It was not known for its quick responsiveness to customers. Inventory managers determined the point below which materials and parts should not fall, and clerks ordered stock whenever it fell below that point. You put a line in a pail of bolts; when you exposed the line, you ordered more bolts. The order was based on average demand, and the fixed trigger point for releasing an order was based on typical behavior.

> **With a pull system alone, the McDonald's next to RFK Stadium would be in trouble.**

These techniques were more or less suitable for managing retail operations, though even in those settings there were problems like the McDonald's franchise on the day of a game. If you ran a garage, for example, it would be tough to restock snow tires quickly after a big blizzard because every tire store would hit its order point and there would be a flood of reorders. The OP, OQ system would not react and place an early order when a blizzard was forecast.

If OP, OQ could be annoying to retailers, it could be disastrous to manufacturers. For want of a part, whole lines were shut down. And job lot orders, when they finally did come in, often forced manufacturers to tie up cash in more stock than they needed. Nor did OP, OQ integrate all information that was available when initiating production: How long does it take to build a product? What is likely external demand? In fact, parts and components are often built for in-house demand rather than for outside customers, and their requirements are well-known.

Managers often took obviously inappropriate action because of the old OP, OQ method. Demand surges invariably caused shortages; small shortages triggered production of parts even when people knew there would be no near-term demand for the part. Against this flawed system, MRP – the first generation of what is now MRP II – had obvious merit and won wide acceptance very quickly.[1]

1. Early HBR coverage included Jeffrey G. Miller and Linda G. Sprague, "Behind the Growth in Materials Requirements Planning," September-October 1975, p. 83.

MRP to the Rescue. The concept behind MRP was straightforward, and MRP II is no real departure. In the same way that the caterer plans and conceives the whole week's production, the MRP system explodes the entire manufacturing operation into discrete parts making up the whole. It then projects demand, the time it would take to meet it, and the materials needed.

The key to MRP is that you have to tell it the lead time to manufacture a part, component, or assembled product. If parts production is intended to support, say, final assembly of a telephone, MRP orders only the parts that are actually going into the phones you expect to sell—not some preset job lot determined by "efficient scale."

So instead of building according to a fixed inventory position of various parts, MRP mandates building to the scheduled delivery of the final product. That, at least, is the theory. The best feature of MRP is its demonstrated capacity to work through the bills-of-materials relationships by which parts and subassemblies become the final product. MRP calculations start at the end items to be shipped and proceed stage by stage through bills of materials, releasing orders for the various parts or assemblies, according to a predetermined quantity and timing. The process then automatically repeats for the next level of parts going into each planned component or assembly.

Penetration of MRP methods into manufacturing has been substantial, especially in industries characterized by complex bills of materials, large numbers of open orders, and many needs for materials coordination among plants, vendors, and customers. Indeed, MRP has become so much the standard for materials management that it has led to the professionalization of the task, as exemplified by the American Production and Inventory Control Society. At the same time, owing to the heavy computer demands of MRP, systems managers and MIS departments have taken over a good deal of manufacturing management.

Push Comes to Shove

MRP II—more exact than MRP before it—initiates production of various components, releases orders, and offsets inventory reductions. MRP II grasps the final product by its parts, orders their delivery to operators, keeps track of inventory positions in all stages of production, and determines what is needed to add to existing inventories. What more could JIT ask?

A major barrier to MRP, though, is the cost of hardware and software for a complex computerized system—no minor barrier especially to smaller producers. And even more important are the costs of training and implementation. You have to teach your workers a lot that they don't know about computers. And in order to enter the right data and the right relationships, you have to spend a great deal of time finding out things about your system that you don't presently know: How should parts be timed to be put together just the way you want them? How long will it take for delivery of all the critical parts?

MRP does not conflict with JIT, but MRP must assume a fixed production environment with fixed lead times. Even with the best intentions, people who set up MRP systems base them on established and often flawed methods of conceiving conversion processes—methods that could be full of inefficiencies and that could be easily improved if workers were not constrained by MRP expectations.

In this context, the professionalization of materials control is a handicap, not a benefit. MRP standards have become a kind of orthodoxy, so people resist the introduction of new methods to the shop floor. New methods can threaten the positions of MIS managers, materials managers, MRP vendors, consultants, and educators who have become attached to the standards.

Of the many standard assumptions made by MRP, the fixed lead times are the most troublesome. Why is MRP so susceptible to getting lead times wrong? The best answer is that production lead times vary depending on the degree of congestion or loading within the shop. The fallacy in MRP is that its releases produce the very conditions that determine lead times, but these lead times have already been taken as known and fixed in making the releases. Consider again the cars traveling to the tunnel. The time it takes for any one car depends on traffic conditions and starting times. Change the pattern of departures, and you change the load on the system and the time it takes.

There is another way of looking at this. A single lead-time number must suffice in MRP for all situations faced on the floor. Consequently, the number has to be set high enough to cover all variations up to the worst case. If an order is ever late, people have an incentive to increase the planned lead time in the system so that the delay does not occur again. A commuter who might encounter an accident or traffic jam will leave early to protect against such contingencies. Similarly, orders will usually be released too early and will often complete early, thereby increasing inventories in the system.

Incentives for Improvement. Perhaps the most pernicious aspect of MRP is the removal of any responsibility for lead-time reduction from the shop floor. How can there be incentives to reduce lead times if there are no rewards for completing work faster than MRP's fixed standards say?

Another big problem with MRP is its unnecessarily complex and centralized nature. MRP II systems plan and coordinate materials flow and produce order releases to the shop floor. But in many situations, the shop floor can be more flexible than MRP II.

For example, an assembly group might want to change its build schedule because parts aren't available for some current schedule. Yet change is stymied because the appropriate paperwork is unavailable and won't be available until the next run of the MRP system–say, next week. It often makes no practical sense to run an MRP plan every day. It takes time to collect and distribute all the data involved. Also, a good-size MRP system can tie up the central computer for hours. The same computer might be used for everything from word processing to payroll and general ledger accounting. Yet some shops would be better off working in just such short cycles.

Some MRP enhancements have addressed these problems. MRP vendors have created "shop floor control" modules–actually monitors, not controllers, which track progress on the shop floor. The resource management tools in MRP II analyze capacity and resource loading. Perhaps the best known of these systems is "rough-cut capacity planning." This method analyzes the load that MRP order releases create on the shop floor. If this load exceeds the capacity of a work center, the implication is that the work in the shop will not get done within the time allowed. The human planner must now find some way to cure the problem diagnosed. Sophisticated techniques for evaluating the lead-time consequences of MRP releases are also available now, including order release methods (X-FLO, for example), scheduling techniques (OPT, CLASS, MIMI), and simulations (FACTOR).

While helpful, these methods increase MRP costs and can be subject to the same criticisms as the system they are meant to restore: they remove responsibility and incentives from the shop floor and they are only as good as the information put into them.

Does Pull Ever Come to Tug?

If MRP superseded OP, OQ, the kanban method is often prescribed as a JIT technique that overcomes the deficiencies of MRP. Presumably, if you set up a production system that works like a bucket brigade, you can forget about providing incentives for continual improvement or gathering what may prove to be incorrect information. The team will discipline itself according to the next customer's needs.

To the extent that kanban works like a bucket brigade, it is indeed a JIT system. Everyone in the chain takes about the same amount of time to pass a bucket, and the system can work without any inventories of buckets between people. If the output end slows down, the whole chain will react and slow down; if it speeds up, the chain will react and speed up as much as possible, until limited by the slowest bucket passer.

Nor is kanban just warmed-over OP, OQ. With kanban systems, workers can see clearly the value of lead-time reduction. Unlike other pull systems, kanban combines production control with inventory control. The interaction between lead times and inventory levels becomes obvious to everybody on the line.

> When MRP punishes workers for completing work early, where are the incentives to reduce lead times?

Moreover, the production supervisor owns the inventories that are produced; they are not pushed into other hands. He or she is thus forced to recognize that increasing the lead time of manufacturing increases WIP as well as finished inventory. This is completely unlike conventional pull systems like OP, OQ in which the inventory management function is separated from production or replenishment.

Indeed, the kanban method of posting circulating work orders makes the current work commitment of the manufacturing cell immediately obvious to everybody in the cell. Planning setups in advance, therefore, or opportunistically consolidating batches to save setups can become routine. The mix changes and demand surges that call for personnel reassignments become more transparent.

Kanban has another virtue that JIT people like. The fixed pool of cards in a kanban cell reduces the extent to which demand fluctuations are passed on by the cell to other upstream cells. The cards provide an upper bound that filters out extreme variations. At the same time, the system disciplines the downstream customer by punishing wide fluctuations or demand surges. A sudden surge will not be satisfied until the limited number of cards circulate many times. This encourages uniform demand and level schedules on the downstream side.

When kanban works like a bucket brigade, it is inherently a JIT system.

Kanban Is Reactive. Kanban is not without difficulties, though, which show up especially when it is forced to operate in complex operations where variations are too great or too intractable to be disciplined easily. Toyota's kanbans discipline suppliers, but a supplier's kanban cannot discipline Toyota.

The kanban method works best where there is a uniform flow – a level-loaded, synchronous, or balanced system. It does *not* plan well. JIT enthusiasts should realize that when a kanban system is implemented in an environment full of variations in supply and demand, it is even less likely than MRP to operate in a stockless manner – that is, without a burdensome amount of WIP. Variability causes the same extreme problems that it does in other pull systems. Extra cards or containers – buffers, for example – have to be introduced to cover variability and avoid back orders. Nothing in a kanban system magically reduces inventory levels due to some internal rule or formula.

Since the system is reactive, changes in demand level percolate slowly from stage to stage. Even if it is perfectly obvious that demand is rising, there is no standard way to prepare for the situation. Some U.S. assemblers that work with Japanese suppliers using pull systems have commented that if there is a steep change in demand levels, the suppliers take from three to six months to adjust to it and encounter plenty of problems until the system reaches smooth operation again.

Tailored Controls, Hybrid Systems

Where does all this leave us? Which system should the manufacturing manager choose? The simple fact is that there is no need to choose between push or pull. These methods are not mutually exclusive, and each has its pros and cons. The best solution is often a hybrid that uses the strengths of both approaches.

Pull methods tend to be cheaper because they do not require computerization – hardware or software. They leave control and responsibility at a local level and offer attractive incentives for lead-time management. MRP systems are good at materials planning and coordination and provide a natural hub for interfunctional communication and data management. When it comes to work release, they are good at computing quantities even if they are weak on timing. A successful hybrid system can use each approach to its best advantage.

The key to tailoring production control lies in understanding how the nature of the production process drives the choice of control method. The accompanying exhibit summarizes various manufacturing control methods and process characteristics. For a continuous-flow process, ongoing materials planning is not essential and JIT supply techniques work well. Order releases do not change from week to week, so a rate-based approach can be used. At the shop floor level, JIT materials-flow discipline combined with pull release – kanban, for example – is effective.

In a repetitive manufacturing environment with fairly stable but varying schedules, materials planning can be a combination of MRP II and JIT methods. Order release may require MRP calculations if changes are frequent or if it is necessary to coordinate with long lead times or complex materials supply and acquisition. Pull methods work well on the shop floor.

As we move to more dynamic, variable contexts – like job shop manufacturing – MRP becomes invaluable for planning and release. Pull techniques cannot cope with increasing demand and lead-time variability. Shop floor control requires higher levels of tracking and scheduling sophistication. Materials flow is too complex for strict JIT.

Finally, in very complex environments, even job release requires sophisticated push methods. Where these are too expensive, the only option is to live with poor time performance, large inventories, and plenty of tracking and expediting.

Tailored Production Controls

Lead-Time Variability (Low → High)	Control Stage: Materials Planning	Control Stage: Order Release	Control Stage: Shop Floor
Pull: Continuous Flow	1 JIT	2 Rate Based	3 JIT-Pull
Hybrid Push-Pull: Batch, Repetitive	4 JIT-MRP	5 Pull or MRP	6 Pull
Hybrid Push-Pull: Batch, Dynamic	7 MRP	8 MRP	9 Pull or Order Scheduling
Push: Custom Engineering	10 MRP	11 Order Scheduling	12 Operation Scheduling

Continuous Flow: The production process is dedicated to one or a few similar products. Production is continuous and level so that the lead time for production is uniform and predictable. Some examples are assembly lines, transfer lines, and dedicated-flow lines.

1. Since production rates are uniform and predictable, material can be delivered to the process in a JIT manner.

2. Work orders are not required since production is level. A blanket order specifying a "going rate" is adequate. Occasionally, if the production mix is changed, the rates may be changed, but these changes are infrequent.

3. The predictability of the process and the production rate make it possible to design for smooth JIT materials flow on the shop floor. If there are points at which small inventories are accumulated for quality control or accounting purposes, they can be replenished in a pull manner.

The author is grateful to Indur Shivdasani for his contribution to this exhibit.

The Best of Both. The dividing line between push and pull is obviously not sharp. In many situations, the two can coexist and are complementary. Most important, it is perfectly possible to take elements of one system and add them to the other. If pull systems have natural lead-time reduction incentives and push systems do not, for example, there is nothing to prevent managers from instituting a program of incentives in the context of a push system. Given the importance of lead-time reduction, in fact, it is crucial for managers to measure lead-time performance and provide feedback on response and turnaround times to each work center and shop. Though MRP systems do little to encourage good lead-time performance directly, managers can introduce measurement and incentive schemes based on MRP's data collection capabilities.

There is nothing to stop managers from compensating for the deficiencies of pull systems either. Pull systems, for instance, have no means of lot tracking–pegging lots to specific customers. But customers may want to keep track of their orders, and there may be special regulatory or quality control reasons for maintaining a lot's identity. So why not add lot tracking and data collection systems to a kanban line, leaving the release function as a pull system? (One

Batch, Repetitive: Parts of the process may resemble a continuous-flow system while others involve multiple products produced in batches. Lead times are fairly constant and predictable. The product mix is relatively constant but may have variations from month to month. Typical is production of parts and components for a high-volume end product–such as cars or electronics.

4. Some parts and materials that are used uniformly can be delivered in a JIT manner. In other cases, with long lead time items, MRP is required to plan purchasing, delivery, and coordination between plants.

5. Since lead times are predictable, MRP works well, but so do pull methods–and they are cheaper. MRP may be required for master scheduling when work orders are generated; inventory must be managed; and work centers must coordinate.

6. Work on the shop floor flows relatively smoothly, and pull systems can be used to move work on the shop floor. If MRP systems are used, the trick is to coordinate pull on the floor with MRP work orders. One simplifying device is to combine several levels of the bill of materials into fewer levels so that the points of coordination with MRP are minimized. Tandem hybrid systems work well.

Batch, Dynamic: Production is in batches, and the output mix and volume can vary; many customers come in with their orders on a weekly and monthly basis. The load on the facility changes; bottlenecks can shift, with backlogs appearing here and there; lead times become variable. Examples are parts and product manufacturers supplying several customers, factories supplying retail outlets with multiple parts, and medium- and low-volume plants.

7. As production mix and volumes change, many different materials and parts are required; departments must coordinate production. MRP becomes essential to match purchasing with production and coordinate parts fabrication and assembly. Production volumes can be smaller than lots likely to be purchased. Inventories build up and must be tracked.

8. Output varies too much for pull systems to work well. Look ahead, and build what will be needed. Even if MRP's timing isn't perfect, it does all the bookkeeping on quantities, inventory availability, and requirements, net of inventories.

9. At the shop floor level, work orders must be tracked. In some early common operations such as metal pressing, blanking, or molding, volumes may be high enough and level enough to use a pull system. Work orders, generating a master schedule, tie together purchasing, parts, subassemblies, assemblies, and customer orders. All are "pegged" and tracked with the MRP system.

Custom Engineering: With low-volume, complex engineered products or with custom manufacturing, there is no regularity in production patterns. The load on the facility can vary widely; what took two weeks when ordered in January might take four months in June. Queues and congestion are a major concern, and lead-time management requires a high level of analysis and detail. Examples of such facilities are machine tool manufacturers, custom-equipment builders, and products with a high option and custom content.

10. There is no regularity in materials usage; some materials may be ordered only after a customer order is received. MRP is invaluable as an information management tool. It books orders, maintains bills, whether custom or standard, and coordinates customer orders, shop orders, and purchasing orders.

11. The factory runs on work orders generated by MRP. But MRP's poor understanding of lead times and capacity limits means that the order releases are of little use for good time and delivery performance. MRP still plays a role, however, in maintaining information about materials and inventory availability and coordination between departments.

12. Scheduling systems (OPT, CLASS, MIMI) that can handle the complexity of detailed operational scheduling are only just appearing. They are too complex and costly for smaller shops.

simple and effective approach is to accumulate the information physically, with the lot itself as it moves through various process stages, and then record it electronically at inventory points in the process.)

Theoretically, there is no limit on the variety of control methods that can be developed. Most are hybrids. Attempts to implement pure push systems are usually accompanied by the growth of some informal, reactive pull procedures. The most common, alas, is the "hot list," by which assembly tells manufacturing what parts it wants most on a given day.

In a way, such informal procedures are only piggybacking on the official MRP system, using short-term release information that MRP has not yet processed. The trouble with any informal procedure, however, is that it is very unsystematic; it may be based on assembly's guess of what it can get from parts and does not take into account the actual position of open orders in parts. Moreover, it undermines the credibility of the official system. Since there can be no coordination between the two, disbelief in the official system becomes self-fulfilling. Instead of such informal overrides of MRP II, consider one of the following hybrids:

JIT-MRP. There are now several modifications of existing MRP II systems, which add pull elements

You tell MRP lead times to make and assemble the parts.

and remove some of the problems connected with the system's lack of responsiveness. Some such modifications are "synchro-MRP," "rate-based MRP II," and "JIT-MRP." These systems are appropriate for continuous-flow or level-repetitive processes, where production is at a level rate and lead times are constant. In these situations, the order release and inventory management functions are of little value. The facility can be designed to operate in a JIT manner so that any material that enters the facility flows along predictable paths and leaves at predictable intervals. Work is released by a pull mechanism, so there is no WIP buildup on the floor.

Such a JIT-MRP line produces to meet a daily or weekly build rate rather than build to specific individual work orders. This means that inventory position isn't necessary for release calculations. Inventory levels can be adequately calculated after the fact on a so-called "back-flush" or "post-deduct" basis by subtracting to allow for production that has already taken place. In short, MRP serves mainly for materials coordination, materials planning, and purchasing and not for releasing orders. The shop floor is operated as a JIT flow system.

Tandem Push-Pull. In a repetitive batch environment where lead times are fairly stable, either an MRP or a pull approach can achieve order release. MRP would be best for purchase planning of items with long lead times. Actual build routines closely correspond with the MRP II schedule, yet the timing of subassembly and assembly releases can be eliminated to allow the shop floor to change rapidly in response to short-term demand pull. Subassembly and assembly are flexible, short-cycle processes that can easily be run on a pull basis.

In this common situation, push and pull systems can simply be juxtaposed – MRP II to ensure parts availability based on end-item schedules and kanban for actual subassembly and assembly releases. MRP can be run only as frequently as necessary for parts purchasing and planning. Since the floor schedules can change quickly, the MRP database will always be playing catch-up with actual part withdrawals. This approach has been particularly successful in subassembly and assembly environments in which manufacturing cycle times are much shorter than parts purchasing and fabrication lead times.

Requirement-driven Kanban. Consider another situation where individual cells within the manufacturing chain can be run with kanban control, although MRP II runs much of the rest of the process. This can occur where final assembly schedules are unstable with respect to volume and mix, yet certain portions of the production process see fairly steady demand. A plastics injection molding cell that makes the same bottle for different shampoos is a good example. The MRP system can predict requirements for plastic parts quite well; kanban could run the injection molding cell.

One approach for such a case is to use MRP II to plan the number of cards in the cell on the basis of the gross requirements for all the parts produced by the cell. The MRP system doesn't have to monitor the inventory level in the cell or match demand with available inventories since the system doesn't make order releases. The gross requirements are an aggregate forecast of demand from the cell. Of course, as the gross requirements increase, additional cards are introduced into the cell in advance of the demand increase. They are withdrawn as the requirement level drops. MRP thus plays the role of planning adviser to the cell, setting the budget level in terms of the number of cards but not specifying the "expenditure" or release of the cards.

Many component manufacturing shops supplying subassembly and assembly operations, where the mix may change substantially but the total volume does not vary much, can use this approach. Other users are builders of common components or subassemblies like motors, similar components like PCBs, and metal-forming operations like blanking, shearing, and pressing.

Dynamic Kanban. Pull methods like OP, OQ typically do have some push component, such as seasonal expectations. Forecasts of demand patterns can be used to set new values for the order quantity and for the order point. In this manner, the otherwise passive pull system is able to anticipate predictable changes.

Similarly, the card quantity in a kanban system can be altered in response to regular changes in demand forecasts–not only seasonal variations but obvious trends or planned promotions. In these cases, the forecast can be used to calculate the number of cards necessary to support the changed level of demand. The cards become a planning parameter driven by forecasts of activity.

Looking Ahead to CIM

There are no panaceas for manufacturing management problems. A single approach will not suffice for all situations. Managers have to design and refine solutions. Kanban itself, like so many JIT techniques, evolved over many years.

I believe that future advances in pull systems will most likely accommodate even more computerized and automated factory environments. The challenge will be to create incentives for process improvements in the automated factory. Expert systems will have important roles in troubleshooting and diagnosing problems, sometimes even substituting for shop floor supervisors.

The fastest growing area for push methods is "factory management systems"–new methods oriented toward shop floor management rather than materials planning. These new systems monitor manufacturing and collect production data and merge with technologies like smart cards and bar coding. Even newer techniques of scheduling and cell management are leading to a bottom-up style of factory management. Indeed, as information technology evolves further, push techniques, like the pull approaches, will tend to decentralize control to the local, cell level.

In today's manufacturing settings, we are witnessing a drift toward the ultimate JIT factory, in which the needs of a JIT cell are perfectly coordinated with the output of all others and matched to customers' varying demands. Expressed in those terms, that's the ultimate CIM factory too.

Reprint 89505

"Tell the court,...is this, or is this not, the carrot your employer dangled in front of you?"

Competing Globally

In Question
Perspectives on business debates-in-progress.

Small business is: a) inherently dynamic, b) too small to survive, c) insignificant to manufacturing, d) none of the above.

Can Small Business Help Countries Compete?

by Robert Howard

For most of this century, it was a common sense of business that success came with size. Then came the 1980s, and this bit of managerial wisdom–like most others–was turned on its head. While bloated corporations were downsizing, small ones were proliferating. According to a much-publicized finding by MIT researcher David Birch, small companies of fewer than 100 employees were responsible for as many as eight out of ten new jobs. When many big companies were slow to respond to shifts in technology and markets, small companies were on the cutting edge of innovation–a trend symbolized by the high-tech startup in California's Silicon Valley and Boston's Route 128.

So what was once a truism–that big is necessarily better–has become the focus of a hotly contested debate. Depending on which side you're on, either small business is inherently dynamic, innovative, and entrepreneurial–in contrast to sluggish corporate bureaucracies–or small business is backward, isolated, starved for R&D, and doomed to extinction in the marketplace or acquisition by powerful global giants.

The problem with this debate is that its images of small and big business are static at a time when both are changing. Neither set of stereotypes is adequate to characterize the diversity and complexity of economic relationships in business today. The question remains: What is the place of small business in a global economy? Put simply: *Can small business help countries compete?*

The books, reports, and other texts discussed here represent a major step forward in the argument about small business. Together, they serve to reframe the debate by making three important–and intriguingly counterintuitive–claims:

1. Small business is most important where it is least predominant.

The very term "small business" evokes images of the service or retail sectors, where small companies are most common. If the issue is international competitiveness, however, the place to look is manufacturing, which, at least for the time being, dominates international trade. Small companies make up a small portion of the U.S. manufacturing sector. But on the scales of international competition, the performance of a small tool-and-die manufacturer or metal-cutting shop weighs more heavily than that of the local shoe store, catering company, or 7-Eleven.

The roughly 355,000 U.S. manufacturers with fewer than 500 employees are responsible for approximately 46% of the value added in U.S. manufactured products. They make up what *TechnEcon*, the newsletter of Michigan's Industrial Technology Institute, calls the "base" of the manufacturing economy. To an important extent, the competitiveness of major U.S. manufacturing corporations depends on these small suppliers. Even highly integrated companies such as General Motors spend more than half of each sales dollar on purchases from suppliers. Cutting in half the in-house production costs at large companies would reduce overall product costs by around 15%. Doing the same at small and midsize companies would double that percentage.

2. The real issue isn't size; it's industrial organization.

According to the traditional view, large size brings efficiency at the price of a certain rigidity, and small

Robert Howard is an associate editor of HBR, where he covers human resource management and industrial organization. His book Brave New Workplace *was published by Viking Penguin in 1985.*

size brings flexibility at the price of instability. But technological and economic developments are making possible a new kind of organization that combines the virtues of both. In this new organizational model, it's not the size of a company that matters so much as the quality of the business relationships tying companies to each other. The key unit of production is no longer the individual company but a decentralized network of companies. Sometimes, these networks consist of vertical links tying small suppliers to large final assemblers. In other cases, the links are horizontal – binding together a number of more or less equal small companies. In both cases, these networks make possible continual innovation through a delicate balance of competition and cooperation, demands and support.

Some examples described in the publications cited here:

□ Japan's supplier-group system, which ties large final assemblers to small suppliers in long-term relationships of collaboration.

□ Europe's industrial districts, where networks of highly specialized small companies combine state-of-the-art technology and skilled labor to produce high value-added products.

□ California's Silicon Valley, where a new generation of semiconductor companies has remained innovative and competitive by constructing complex alliances and cooperative relationships.

3. A country's capacity to build strong production networks constitutes a new kind of competitive advantage.

These new business relationships do not develop automatically. It takes managers willing to redefine how they interact with suppliers, customers, and competitors. And it requires government, trade associations, unions, and other social groups to create an institutional context – think of it as a new kind of infrastructure, as important as highways or a telecommunications system – in which networks can flourish. Countries that successfully establish such an infrastructure will have an advantage in the international marketplace that others will not.

TechnEcon: Research Newsletter of Automation and American Manufacturing
Ann Arbor, Michigan: Industrial Technology Institute
Published three times per year.

The Re-emergence of Small Enterprise: Industrial Restructuring in Industrialized Economies
edited by Werner Sengenberger, Gary Loveman, and Michael Piore
Geneva, Switzerland: International Labour Organization, 1990
312 pages.

"Skills Without a Place: The Reorganization of the Corporation and the Experience of Work"
by Charles Sabel
An address to the plenary session of the British Sociological Association
April 2, 1990.

Making Things Better: Competing in Manufacturing
A report of the Office of Technology Assessment, U.S. Congress
Washington, D.C.: Government Printing Office, 1990
244 pages.

The Machine That Changed the World
by James P. Womack, Daniel T. Jones, and Daniel Roos
New York: Rawson Associates, 1990
323 pages.

Flexible Manufacturing Networks: Cooperation for Competitiveness in a Global Economy
by C. Richard Hatch
Washington, D.C.: Corporation for Enterprise Development, 1988
27 pages.

"Regional Networks and the Resurgence of Silicon Valley"
by AnnaLee Saxenian
California Management Review 33, No. 1, Fall 1990.

Employers Large and Small
by Charles Brown, James Hamilton, and James Medoff
Cambridge: Harvard University Press, 1990
109 pages.

These three claims offer a fresh way to think about small business and the competitiveness debate. The relative lack in the United States of small-business networks and the infrastructure to support them goes a long way toward explaining why, with a few exceptions, the small-manufacturing sector in the United States lags behind that of other major industrial nations. Ironically, while many in the United States have been celebrating small business, other countries – in particular, Japan and West Germany – have been using public policy to make small manufacturing a powerful asset of international competition.

> After declining for decades, small business's share of employment is growing.

The rediscovery of small business in the 1980s quite rightly emphasized the creative energies of individual entrepreneurs and the benefits of relatively unconstrained competition. Building a competitive small-business manufacturing sector in the 1990s may require something else again: an understanding of the economic benefits of cooperation and the importance of public policy to reshape business relationships to the ends of competitive success.

The Business Logic of Small-Scale Production

To understand the new role of small business in manufacturing, begin with a dramatic empirical finding: for most of this century, the size of the small-business sector (measured by the share of overall employment in companies with fewer than 100 employees) was shrinking throughout the industrialized world. But starting around 1970, that trend began to reverse. Small business's employment share began to grow slightly. When small units of large businesses are added to the calculation, the trend toward small-scale production is even more pronounced.

This is the central conclusion of *The Re-emergence of Small Enterprise,* a major international study of small business funded by the International Labour Organization (ILO). The study analyzes the role of small business and its share of employment in nine countries, including the United States, West Germany, and Japan. According to the volume's editors – economists Werner Sengenberger of the ILO, Gary Loveman of the Harvard Business School, and Michael Piore of MIT – growth in industries like services, where small companies are in the large majority, explains only part of the increase. The employment shift is also happening within specific sectors. For example, between 1980 and 1986, manufacturing employment at large U.S. companies of 500 employees or more declined by 10.8% – nearly 1.8 million jobs. During the same period, small manufacturers employing less than 100 workers added 326,000 jobs – an increase of 7.5%.

For Sengenberger, Loveman, and Piore, what's important about this trend is not the size of the shift – in most cases, it is small, a matter of a few percentage points. The fact that it is occurring so uniformly throughout the industrialized world – at more or less the same time and in many different countries with widely varying economic structures – suggests that something fundamental is going on. Employment growth in both small companies and small units of big companies reflects basic economic and technological changes. In the early part of this century, standardized mass markets and rigid mass-production technologies gave rise to the large, vertically integrated corporation. Today, increased demand for more specialized products, nonstop technological innovation, and cheaper and more flexible computer-based production technologies are inaugurating a decentralization of production.

In their 1984 book, *The Second Industrial Divide,* Piore and MIT political scientist Charles Sabel coined the term "flexible specialization" to describe this phenomenon. In a recent address to the British Sociological Association, Sabel described the new kind of organization that is emerging in the era of innovation-based flexible competition. His suggestive term for it is the "metacorporation" – a work organization "designed to be easily redesigned."

Managers hardly need to be reminded of the imperatives driving this trend. In a volatile business environment in which a company's competitive advantage can disappear almost overnight, the chief managerial challenge is to reduce the costs and – even more important – the time of product development. The faster a company can conceive, design, and bring new products to market, the more successfully it will incorporate innovations and shift rapidly from declining market segments to more prosperous ones.

The precondition of this competitive strategy is a radical increase in organizational flexibility. To speed the development process, companies must link the conception of a product more closely to its actual manufacture in a process of "simultaneous engineering," which blurs the hierarchical distinctions within a company. Traditional divisions between R&D, engineering, manufacturing, and sales give way to product-development teams.

But Sabel goes on to argue that this blurring of boundaries takes place not only inside the corporation but outside it as well in its relations with

CARTOON BY GEORGE DOLE

other companies. In the traditional mass-production organization, for example, supplier relations are at arm's length and adversarial. Customers maintain a large vendor base, frequently shift orders among vendors in a search for the cheapest price, and favor short-term contracts, rarely longer than a year. But to take advantage of the latest innovations and speed the design and production of components, relations with subcontractors must become much more collaborative. Large corporations jointly design new components with their suppliers, enter into joint ventures to exploit a new technology, or spin off technologies difficult to commercialize in-house to new startups. As a result, "the corporation becomes more a federation of companies than a single organizational entity," tied together in a system that Sabel terms "collaborative manufacturing."

What's more, the loosening of boundaries extends to the relationships between business and other social institutions. As decentralized collaborative manufacturing spreads, dense regional clusters of highly specialized small units flourish. Too small to do their own R&D, worker training, and the like, these small companies come to depend on outside institutions – universities and community colleges, trade associations, governments – to provide these services collectively. Thus the metacorporation is woven into the social fabric far more tightly than the vertically integrated corporation of the past.

These trends create a new economic space for a particular kind of small company that is close to the market, technologically dynamic, and linked to other companies in complex networks of production. The ILO study identifies two types of networks: "kingdoms" tie small suppliers to a large corporate customer in a vertical supplier chain, under the strategic direction of the big company; "republics" join small, highly specialized companies to each other in a horizontal network where no one company dominates. These two models are not mutually exclusive. As befits the logic of flexibility, they are highly fluid. The same company can shift from one to another over a period of time or even belong to both simultaneously.

What do these networks look like? Consider three examples: a classic kingdom, the Japanese supplier-group system; a classic republic, Europe's industrial districts; and a hybrid that seems to transcend both, second-generation semiconductor companies of Silicon Valley.

Kingdoms and Republics

The predominance of small business in the Japanese economy is well known. About 75% of manufacturing employment in Japan is in small and midsize companies, in contrast to about 35% in the United States. Less well known is the central role these small suppliers play in providing the flexibility and high quality that makes Japan's large corporations such formidable international competitors.

Both *Making Things Better*, a report of the U.S. Congress's Office of Technology Assessment (OTA), and *The Machine That Changed the World*, the product of a five-year study by the International Motor Vehicle Program at MIT, describe this role in admirable detail. In auto, electronics, and other key manufacturing industries in Japan, production is organized according to the supplier-group system – a network of large and small companies arranged in a pyramid. At the top, a large "parent" company is responsible for final assembly. It deals with a first tier of smaller companies who make major components and subassemblies. These first-tier suppliers then manage relations with a more numerous second tier of manufacturing specialists. These companies in turn work with a third and sometimes even fourth tier of smaller and smaller companies specializing in increasingly narrow tasks.

Unlike traditional customer-supplier relations in the United States, Japanese suppliers have long-term relationships with their customers, sometimes lasting over decades. According to one recent survey of Japanese subcontractors, 68% of the respondents had never changed their parent, and 53% had been doing business with the same parent for 15 years or more. Even more important, Japanese suppliers and customers constantly share detailed information about production costs and techniques. And joint agreements about how to share savings from cost reductions and productivity improvements are made before a particular job is undertaken.

Most descriptions of the supplier-group system present it as little more than a way for powerful, big companies to take advantage of weaker, smaller ones – for example, by forcing suppliers to bear the costs of downturns. As both the OTA report and the MIT study rightly emphasize, this is much too simplistic a view.

Lead companies do put great pressure on their suppliers. They insist on high quality and continual price reductions. And since they typically have more than one source for a particular component or part, they can punish a supplier who fails to meet these demands by cutting its share of sales in favor of a more compliant competitor. What's more, wages at subcontractors, especially in the lower tiers, are at least 25% below those at the lead companies, and workers do not enjoy the employment security that the core workers at big companies do. But lead companies also provide a great deal of support, primarily financial and technical, that allows the small suppliers to continually upgrade their technology and production techniques.

> Japanese companies both pressure and support their suppliers.

It is precisely this combination of competition and cooperation that makes for a responsive, high-quality, and increasingly dynamic industrial system. The supplier-group system allows the lead company to stay lean and focused. The responsibility for designing and manufacturing whole subassemblies is delegated to first-tier suppliers. Nissan, for example, has only one seating supplier for its new Infiniti Q45 model. General Motors, in many cases, is still dealing with 25 suppliers providing the 25 needed parts to the seat-building department of its assembly plants. The strong ties and frequent informal contacts between customers and suppliers also make simultaneous engineering of new products easier. Japanese suppliers often take part in the joint engineering and analysis of new components up to two years before a new product is manufactured. In 1988, Nisshin Kogyo, a leading Japanese brake manufacturer, had a product-development team of seven engineers, two cost analysts, and a liaison person located at Honda's R&D center to work daily with Honda's development engineers on the design of a new car.

> Nissan has one seating supplier for its new Infiniti Q45; GM has as many as 25.

Small companies also benefit. They get reliable long-term markets. Even more important, the mix of competition and cooperation is a constant spur to innovation. In automobiles and electronics, parent companies lend state-of-the-art equipment to suppliers and sometimes even provide the financial help to buy it. They also assign engineers or technicians to teach small companies how to use new equipment or organize work more efficiently. In this way, new technologies cascade through the system. And as small companies become more adept at the use of these technologies, the flow of assistance and innovation begins to move up as well as down – spreading the risks of innovation throughout the system and creating a huge outside source of product and process improvements for the parent company.

But close ties between big and small companies are not the only source of the competitiveness of Japan's small manufacturing companies. Government policies and aid have also been decisive. The OTA estimates that the Japanese government provides about *20 times* more financial aid to small business than the U.S. government does. In 1988 alone, low-cost direct loans to smaller companies from government financial institutions in Japan amounted to more than $27 billion – with loan guarantees providing an additional $56 billion.

The Japanese government also provides half the $470 million annual budget for a national system of 185 testing and research centers. For a small fee, companies can use equipment at the center that they can't afford to purchase on their own and consult staff engineers about special problems. This "dense, nationwide network of free, public technology extension services for small and medium-sized firms," argues the OTA, is a second crucial source of support for small Japanese companies.

In contrast to kingdoms, republics feature horizontal connections between specialty producers that work together to produce complete components and finished goods. Strategic direction is provided not by one big company but by a "broker" – usually a public institution or private-membership organization – that delivers the services and coordination individual companies cannot provide on their own.

The classic example of the republic model is the industrial districts of Europe. In Italy, West Germany, and Denmark, entire regions have developed thriving manufacturing sectors based on this network model. C. Richard Hatch's *Flexible Manufacturing Networks* is a good introduction to the phenomenon.

Consider, for example, the region of Emilia-Romagna in north central Italy. With a population of only 4 million, Emilia-Romagna has some 90,000 small manufacturing companies, the vast majority with 50 employees or less. These companies account for about 40% of the region's employment and span a variety of industries, including machine tools, automatic machinery, motorcycles, automobiles, electronic controls, and apparel.

Although manufacturing businesses in Emilia-Romagna are tiny, through networks they can compete globally in state-of-the-art technology niches. Take the Modena-based robotics manufacturer FSM. With only 16 workers, it is at the center of a small network that produces globally competitive robotics for the diesel engine industry. FSM handles the systems design, assembly and testing, and global marketing. Meanwhile, five other local companies supply the electronic controls, machining, hydraulic components, welding, and fabrication. None of these companies has more than 20 employees. Only one big company is involved in the network: the German electronics giant Siemens provides the motor.

Like the Japanese supplier-group system, the horizontal networks of Emilia-Romagna maintain a creative tension between cooperation and competition. Small companies compete to join the most successful and most profitable networks. And companies frequently belong to many different networks at the same time – for example, serving as suppliers to large companies even as they pursue joint development and production with other small partners. The republic structure also allows them to cope with rapid changes in technology and markets. According to Hatch, one reason FSM does not simply buy out its partners is that given the uncertainties of technological development in robotics, a certain measure of independence is preferable to integration.

> FSM is a globally competitive robotics manufacturer; it has only 16 employees.

Government assistance has been crucial to the development of the European industrial districts as well – although in a different way than in Japan. The networks in Emilia-Romagna are largely the creation of the regional government, in collaboration with entrepreneurs organized in trade associations and workers in unions. For example, the Emilia-Romagna government has helped fund ten applied R&D centers for networks throughout the region. One, known as CITER, does market forecasting, performs design research, and provides access to new technologies for some 600 small knitwear companies in the city of Carpi, the center of Emilia-Romagna's high-fashion knitwear industry. Through CITER, these companies have access to advanced technologies and global markets they could never afford to develop on their own. While CITER's government funding was temporary – today the organization subsists entirely on membership dues – the funding was a crucial catalyst in forming the network.

In both Japan and Italy, these new business relationships linking companies more closely to each other and to other social institutions have transformed what traditionally has been a relatively backward sector of the economy. Today Japanese small manufacturing companies increasingly keep pace with big ones in the purchase of advanced technology. As they improve their technological capability, some first-tier suppliers are able to take on more complex tasks, including the assembly of entire products, and pass on their previous tasks to second- and third-tier suppliers, allowing them to move up the supplier chain as well. And while, for the most part, wages in Japanese small manufacturing

CARTOON BY H. MARTIN

still lag behind those at big companies, the best small companies pay even more than the large corporations do.

In some cases, small companies are even breaking the exclusive bonds that have traditionally tied them to a single parent. For example, in the Tokyo industrial district of Ota-ku, famous for its thousands of innovative small factories and where some 95% of the roughly 9,000 plants have 30 or fewer employees, there are now about 1,000 companies with no strong links to major corporations. The government is encouraging such companies to form cooperative associations that replace the vertical ties to big companies with horizontal ties to other similar small companies. These networks of small companies work together on R&D and share technical, management, and marketing information. In this respect, at least, it would appear that some Japanese kingdoms are turning into republics.

In Emilia-Romagna, collaborative manufacturing has been an engine of economic development. In 1970, Emilia-Romagna was the fourth poorest of Italy's 21 regions, its industries buffeted by growing international competition. By 1985, it was the second wealthiest, and wage rates were 25% above the Italian average. In fact, some networks have become so successful that they have become the target of takeovers by large corporations. According to the OTA report, large companies are coming to play a more dominant role in some Italian industrial districts. In some cases, decentralized production continues, but under the growing financial and strategic control of locally dominant companies or outside corporations. If kingdoms can turn into republics, republics may also be turning into kingdoms.

And as a third example of a manufacturing network suggests, new hybrids are growing that seem to explode the distinction between kingdom and republic altogether.

Revisiting Silicon Valley

Silicon Valley's high-tech startups were the prime symbol of small-business dynamism in the 1980s–that is, until the Valley's commodity semiconductor manufacturers ran into the wall of Japanese competition. According to an article by AnnaLee Saxenian in a recent issue of the *California Management Review*, a second generation of small semiconductor companies in Silicon Valley has so far avoided the mistakes of its forebears by creating a third version of the network model.

Companies like Cypress Semiconductor, Weitek, and Altera are small specialty chip designers and manufacturers. By focusing on short runs of high-performance components and custom products targeted at niche markets, these companies are consistently able to stay at the forefront of the innovation process and introduce new products far more quickly than traditional semiconductor companies. They do so by establishing strong horizontal ties to each other–much like republics. Since 1979, new semiconductor startups in the region have forged more than 350 alliances with each other and with other companies. Most involve technology sharing, subcontracting of chip fabrication, or joint product development. Like kingdoms, the ties between these companies also extend to large corporations–both first-generation semiconductor manufacturers and computer systems makers. But in no sense do these big companies dominate the network. Indeed, in many respects, exactly the opposite is the case.

> In Silicon Valley, small companies cooperate with big companies–without losing their room to maneuver.

Take, for example, two small companies: Weitek designs high-performance chips for math-intensive operations. It has 230 employees and did $49 million in sales in 1989 with a net income of about $7 million. Altera makes programmable logic devices. With 300 employees, it had around $60 million in sales and a net of $11 million.

Neither Weitek nor Altera has any chip-production facilities. Instead, they have formed strong cooperative relationships with a variety of small and large chip and computer systems manufacturers. For example, Altera has a $7.4 million equity stake in Cypress and has agreed to go as high as $15 million. In exchange, the company receives a guaranteed fraction of Cypress's output at cost plus, and early access to the company's next-generation manufacturing technology. Meanwhile, Cypress gets an infusion of cash, the opportunity to produce at capacity, and a right to Altera's state-of-the-art products. Weitek has formed similar tight relationships with suppliers and customers–most visibly, an agreement with systems maker Hewlett-Packard to manufacturer Weitek chips in HP's own state-of-the-art fabrication facility.

Weitek is a small company and HP a large one. But in no sense is HP Weitek's "parent." Even in their relationships with big corporations, the small Silicon Valley companies' technological expertise gives them far more room to maneuver than suppliers typically have. Part of the agreement is that the Weitek chips manufactured at HP can be sold to anyone–including HP competitors like Sun Microsystems. And in exchange for giving Texas Instruments a license to some of Altera's second-generation products, the company has won the right to use some of TI's manufacturing processes at other semiconductor manufacturers.

As these last examples suggest, cooperation among Silicon Valley semiconductor companies does not lead to the kind of tight integration that hinders innovation. Much like the Japanese supplier groups and European industrial districts, cooperation is always spiced by intense competition. Both customers and suppliers make an explicit effort to avoid dependence on any one company and to preserve their own autonomy. For example, most Silicon Valley companies prefer that no single customer account for more than 20% of a supplier's output.

The result is a system that allows companies to share the costs and the risks of innovation. Relationships leverage a company's presence in the marketplace, minimize fixed costs, and get products to market faster. According to Saxenian, product-introduction times have shrunk from more than two years to as little as nine months. The success of these companies, she writes, "demonstrates that neither scale nor vertical integration is necessary to survive in the increasingly capital-intensive semiconductor industry."

Government doesn't play the role in Silicon Valley that it does in either the Japanese or Italian examples. Indeed, Silicon Valley entrepreneurs are among the most antigovernment businesspeople around. And yet, Saxenian emphasizes the importance of regional institutions – Stanford University, trade associations, local business organizations, and specialized consulting, market research, PR, and venture capital firms – to the small semiconductor companies' success. These institutions are the glue that holds Valley networks together, providing a variety of services that allow companies to focus on their distinctive competence.

U.S. small manufacturing companies consistently lag in implementing new technologies.

Saxenian argues that the reason first-generation semiconductor companies like Advanced Micro Devices or National Semiconductor have had such problems competing in the global economy is that they abandoned this cooperative infrastructure. As these companies grew, they became increasingly integrated manufacturers of commodity products. And following traditional American mass production, their relationships with suppliers and customers became more adversarial.

The second-generation companies are a response to this organizational choice by the commodity producers. But to succeed in the long term, Saxenian argues, they will have to institutionalize their cooperation by creating their own version of the Italian service centers to supply the worker training, long-term R&D, and strategic planning that no one company can supply on its own.

The New Rules of Small-Business Competition

These three diverse examples offer some common lessons. Call them the new rules of small-business competition: Forge close relationships with customers, suppliers, and other partners. Maintain a creative tension between competition and cooperation. Focus on continual innovation. And create the kind of social infrastructure that both pushes and helps small companies to achieve these goals. How does U.S. small manufacturing stack up to these criteria? Despite the Silicon Valley case, not very well.

Certainly, the best U.S. manufacturers – companies like Motorola or Xerox come to mind – have made enormous strides in recent years toward improving the quality of their supplier networks. They have realized that, just as a company's human resources help define its distinctive core competence, so does the quality of its relationships with small suppliers. They have reduced the number of suppliers, lengthened contracts, collaborated more closely in product design, and redefined the customer-supplier relationship as a partnership. And, of course, many small companies cooperate informally with each other, even though they don't think of themselves as belonging to a formal network.

Still, the evidence marshaled in these studies suggests that many U.S. small manufacturing companies suffer from some weaknesses. They consistently lag far behind big companies in implementing the new technologies on which increased productivity depends. Just to give one example: in the largest survey to date on the rate at which U.S. companies are adopting new production technologies (conducted by the U.S. Department of Commerce), large plants with more than 500 employees were as much as 16 times more likely to use certain technologies than small plants with fewer than 100 workers.

What's more, even as supplier relations are changing, the legacy of the traditional adversarial model looms large. In some cases, large companies use longer term relationships with suppliers merely as a way to shift costs and competition to smaller companies – in effect, exerting pressure without offering support. Often the only change has been to shift the adversarial relationship from the traditional criterion – low price – to new criteria such as quality or on-time delivery. And large customers still rarely influence small-company decisions to use new technologies – let alone provide the kind of financial and technical assistance that is common in Japan. The authors of *The Machine That Changed the World*, for example, argue that in the U.S. auto industry, "the reforms made to date have involved pushing the traditional mass-supply system to its limits under pressure, rather than fundamentally changing the way the system works."

But the greatest weakness of U.S. small manufacturing may not be in management but in politics. In the all-important area of small-business economic policy that shapes the institutional framework in which small companies operate, outdated assumptions hamper effective action.

To the degree that small businesses in this country have a political agenda, it is to exempt themselves from government regulations such as plant-closing legislation, mandated health benefits, or increases in the minimum wage. John Motley, chief lobbyist for the National Federation of Independent Business, recently expressed the typical view: "The vast majority of our members don't want anything from government except to be left alone."

Predictably, such attitudes have ensured that federal small-business policy is both minimal and misdirected. Aside from some small programs for businesses owned by minorities, women, and people with disabilities, the U.S. Small Business Administration makes no direct loans to small business. The United States has nothing to compare to Japan's nationwide network of technical assistance to small companies.

In the absence of coordinated federal action, some state governments are actively encouraging small-business economic development. Although Birch's claim that small business creates the vast majority of new jobs has been largely discredited, most of these programs still focus more on generating employment than on improving the competitiveness of existing companies. By 1988, 43 states were providing some $700 million in startup financing to new small businesses. In contrast, total funding of state *and* federal programs for the technological modernization of already existing manufacturing companies was less than a tenth as much–roughly $40 million to $50 million per year.

> In West Germany, government policies that don't favor small business actually help it to compete.

The problem is that in the absence of a coherent small-business competitiveness policy, it is too easy for small companies to compete on low wages rather than high quality and productivity, thus dooming the small-manufacturing sector to second-class status. As economists Charles Brown, James Hamilton, and James Medoff point out in *Employers Large and Small*, not only do small companies not create anywhere near as many jobs as their advocates have claimed, they also pay lower wages (by as much as 30%), provide fewer benefits, and in many cases offer poorer working conditions.

Given the new rules of global competition, this makes neither social nor economic sense. Consider, by contrast, the approach in West Germany, where small and midsize companies–known as the *Mittelständ*–combine state-of-the-art technology and skilled labor to produce high value-added products. Mittelständ businesses dominate the machinery and machine-tools sector, the most powerful in the German economy, and play a central role in others as well. In chemicals, for example, corporate giants BASF, Bayer, and Hoechst produce only one-quarter of the industry's output. Another 1,600 companies–most with fewer than 500 employees–account for the rest.

In West Germany, unlike the United States and Japan, there is almost no gap in wages and working conditions between Mittelständ companies and large corporations. The reason is the institutional factors that shape the environment in which the Mittelständ companies operate. Government policies place stringent demands on companies, then offer them effective support in meeting the demands. For example, German labor law mandates sector-wide collective bargaining and a high minimum wage. This substantially narrows the wage gap between small and large companies. At the same time, West Germany's excellent national vocational-education system ensures that small companies have access to highly skilled labor. The result is a system that forces small business to compete on high quality rather than low wages.

This and other foreign examples suggest a new direction for U.S. small business. Call it "public entrepreneurship," in which the most important "startups" are organizations that help small manufacturing companies adjust to the new competitive environment.

Some interesting initiatives move in that direction:

☐ The Michigan Modernization Service is a state office that provides assistance with technology, marketing, and worker training to small manufacturers in Michigan. The state government has also helped fund the Industrial Technology Institute, an applied research and consulting organization for local companies. One of its programs works to strengthen the traditionally weak ties between auto-industry suppliers by identifying local companies with complementary production capacities and encouraging collaboration.

☐ The state of Pennsylvania's new Manufacturing Innovation Networks program (MAIN) provides money for small companies, trade associations, unions, and other groups engaged in cooperative efforts to solve shared business problems.

☐ The U.S. National Institute of Standards and Technologies (NIST) has established three regional national manufacturing technology centers. With an annual budget of $9 million, their mission is to work with small and midsize companies to improve their ability to assimilate new technologies. Through NIST programs, some companies are experimenting with the network models that have proved so successful in other countries.

Such initiatives are only a beginning, and it remains to be seen what combination of new business relationships and public support will best suit U.S. small manufacturing. But at least one thing is clear: small business can help countries compete only if countries create the conditions that make for a dynamic small-business sector. In the end, a society gets the kind of small business it wants.

Reprint 90604

Four Corners

Business insights that cross borders.

Welcome to the real world of Japanese manufacturing.

The Feudal World of Japanese Manufacturing

by Kuniyasu Sakai

Years ago, while visiting friends in the United States, I happened to see the classic children's film *The Wizard of Oz.* Near the end, Dorothy at last appears before the all-powerful wizard–a terrifying image of smoke and light. Only when her dog, Toto, tugs at a curtain over to one side does she see a very ordinary-looking man operating a variety of levers and switches to create the wizard's fearsome image. The voice in the smoke bellows, "Pay no attention to the man behind the curtain," but Dorothy persists. In an instant she has shattered the myth, revealing the truth about the wizard and his technique for perpetuating his power.

Kuniyasu Sakai is cofounder and former chairman of Taiyo Kogyo K.K., the parent company of the current Bunsha Group. He is author of many books, incuding Bunsha *(Asahi Shimbun-sha, 1986), which is available in eight languages.*

Why, I ask myself, don't U.S. companies do the same thing with the mighty industrial combines of Japan?

Whenever I travel abroad these days, I make a point of perusing local bookstores. I am always amused to see books like *The Secret of MITI* or *The Key to Japanese Management* or some such thing. I suppose these books are intended to explain the myth of "Japan, Inc." to Western executives who are stymied by the allegedly inscrutable nature of our Oriental business practices. And I suppose that, as a Japanese business executive, I should be flattered that so much attention is paid to our forms of business and management. Yet as I flip through books that explain everything from how to exchange business cards to the structure of the Japanese government, I wonder how much real use any of this is to businesspeople who still don't know the basic facts about the companies they are going to visit.

I say this because, in my conversations with Americans and other foreign businesspeople, I am constantly amazed at how little they seem to know about the realities of Japanese industry. At a time when Japan accounts for 15% of the global economy and Japanese executives are busily studying U.S. and European industry, it seems both foolish–and in some ways dangerous–for Western executives to have such a tenuous understanding of their Japanese trading partners.

Over the past four decades, I have built up a group of several dozen small and midsize companies in a wide variety of businesses, most related to electronics manufacturing. I know well the real world of Japanese manufacturing. I also know that foreign executives have no idea how it works.

My businesses produce high-tech products for some of the best known companies in Japan–all of which are familiar to customers around the world. Yet my companies' names remain unknown, as they should. They exist to support the efforts of the larger companies that can afford to advertise and distribute the products we make. I am happy to leave that business to them. I do not even mind that customers the world over buy products that one of my companies designed and built, all the time praising some famous Japanese company whose logo is on the switch. That is the nature of my business, and I really do not expect consumers to know or even care who built their television or their computer.

I do, however, expect Western executives in important manufacturing industries to know. And what astounds me–and my Japanese colleagues–is that, despite so much "revisionist" thinking about Japan, so many Western businesspeople still know so little about corporate Japan. It seems that the myths of the 1960s are still alive and well. The most prominent and enduring of these myths is the notion that Japanese industry is made up of a handful of powerful giants with factories spanning the nation and workers forming an army of loyal employees who are cared for until retirement by a paternalistic corporation.

This is absolute nonsense.

The Secret Revealed

The giant Japanese manufacturers have become household names worldwide. Companies like Matsushita, Toshiba, NEC, Hitachi, Sony, and Fujitsu have become strong because they produce what the world wants to buy. Their reputations for advanced R&D, innovative products, low-cost and high-quality manufacturing are legendary. Moreover, they seem to have an uncanny ability not just to invent remarkable new products but also to borrow ideas, rework them, tinker with them, and produce something totally "new" from a product concept that originated elsewhere. More often than not, they shrink the new product, add a few gadgets, and find a way to sell it for half what the industry was expecting. A year later, they bring out a newer model and cut prices on the old one before anyone else even has a copy on the market.

How do they do that? What is their secret?

How can the giant Japanese manufacturers, big and resourceful as they are, continue to come up with one idea after another, make the quantum jump from theoretical to applied technology in a single year, and squeeze production costs below what should be economically feasible? And how do they do it year after year, growing more profitable at every step along the way?

> How do the giant companies come up with new ideas, bring them to market so fast, and cut costs? Simple: They don't.

The answer is simple: they don't.

Like the wizard of Oz, Japan's giant industrial combines are not what they appear to be. They do not develop all of their own product line, nor do they manufacture it. In reality, these huge businesses are more like "trading companies." That is, rather than design and manufacture their own goods, they actually coordinate a complex design and manufacturing process that involves thousands of smaller companies. The goods you buy with a famous maker's name inscribed on the case are seldom the product of that company's factory–and often not even the product of its own research. Someone else designed it, someone else put it together, someone stuck it in a box with the famous maker's name on it and then shipped it to its distributors.

Does this operation sound unnecessarily complex? Obviously, these huge corporations have their own factories and workers. So why don't they employ their own resources to produce the goods they sell?

They do, of course–but only partially. For instance, it would make very little sense for an electronics giant like Matsushita to farm out the design, manufacture, and assembly of a refrigerator or microwave oven. These products are ideally suited to mass production in the kind of large, highly automated factories that the giant companies can afford. Their factories produce hundreds of thousands of these units every year.

But what about products that companies must continually redesign to compete for public acceptance–like headphone stereos, small compact disk players, or personal computers? Redesigning means retooling a production line. It means sourcing new parts and lots of other things. For a typical product, a company might expect to sell 30,000 units in a few months, retool, sell another 50,000 units, redesign some basic components, retool again, see what the competition brings out, retool again, and on and on, throughout the life cycle of the entire product line. Although some of the giant makers are now employing the newest flexible manufacturing systems (FMS) to allow them more freedom in production, this retooling process is something many big companies want to eliminate.

Thus they farm out much of this business to subcontractors–smaller companies they can depend on. These companies in turn, faced with redesigning and producing a product three or four times a year, will subcontract the design or manufacture of a dozen key components to still smaller companies.

How extensive is this subcontracting pyramid? Would you guess a few dozen companies? A few hundred? Think again. One electronics company I know has well over 6,000 subcontractors in its industrial group, most of them tiny shops that exist just to fill a few little orders for the companies above them.

Welcome to the real world of Japanese manufacturing.

Dai Kigyo and Chu-sho Kigyo

When Western executives talk about "Japanese industry," they are almost always referring to the big, famous companies–generally the top 1,000 or so that are listed on the first section of the Tokyo Stock Exchange. To all outward appearances, these businesses are the backbone of Japanese industry, the pillars on which the Japanese economy rests–which is exactly what these large companies would like you to believe. But the reality is quite different. Small to medium-size companies, *chu-sho kigyo*, make up more than 99% of Japanese industry and are the real foundation of the Japanese economy. The large companies, *dai kigyo*, rest at the very top of a huge corporate pyramid.

According to Ministry of Finance statistics, in 1988 there were roughly two million registered joint-stock corporations in Japan (there are many more unregistered family businesses). Among the official Ministry of Finance-registered businesses, more than 600,000, or 30%, were capitalized at less than $14,000. Roughly another 30% were capitalized between $14,000 and $36,000, and another 15% at less than $70,000. In other words, over 75% of all registered Japanese companies are capitalized under $70,000–not what you would call major industry.

In fact, only 1% of all companies in Japan are capitalized over $700,000, and less than half of those qualify as large-scale businesses, according to the Ministry of International Trade and Industry. Yet these are the only Japanese companies you ever hear about. Too many people are attracted by the glamour of the giant companies and miss the other 99% of Japanese industry. But it is a simple fact that Japan's internationally famous companies would not exist without the support of literally millions of unknown small businesses spread across Japan.

Why is this basic truth so little known outside Japan? How could so many Western writers who study Japanese industry fail to understand its underlying structure? I doubt that they are all poor investigators. Rather, I suspect they don't see the truth because the big Japanese companies, where they go to do their research, don't want them to know. When I meet Western scholars in Japan, I ask them what kind of factories they have seen. Inevitably, they answer that the big companies have shown them around–which virtually guarantees that they have never seen or even been told of the thousands upon thousands of little shops that keep the giant companies in business. After all, the big companies have nothing to gain by introducing outsiders–especially Western writers–to the inner workings of the Japanese subcontracting machinery.

> From the day a subcontractor accepts its first contract, it has lost its freedom.

According to my Japanese dictionary, the word for "subcontractor" is *shitauke*–pronounced sh'ta-ookay. But from talking to Western manufacturers, I'm convinced that something is lacking in the translation. In the West, as I understand it, subcontractors are free agents. Basically, they can work when they want and for whom they want. If they develop a good relationship with one company and want to continue working with it, that's their privilege. If a company treats them badly, they can send an invoice when the contract is finished and say good-bye. They are independent.

Not so in Japan. From the day a subcontractor accepts the first contract–probably from a small subsidiary of one of the giant companies–it has given up its freedom. It is told what to make, when to put it on line, and how much it will get for it on delivery. If the company that placed the order feels a profit squeeze, it can easily order the subcontractor to reduce its final price. If hard times continue, the larger company can demand yet another cut. If it gets to the point that the subcontractor is losing money on each unit it's producing and has cut expenses and streamlined production to the utmost, the "parent" company could demand that it buy some new piece of equipment to increase productivity. And even if the subcontractor neither needs nor wants the equipment, it has no choice: if it refused, the flow of orders from the parent would dry up overnight–and its business would be gone. Of course, the parent would remain unaffected; it can always obtain supplies from any of the scores of other companies in its group.

I know something about the tension between small companies and the big *keiretsu* or industrial groups that swallow them up, as a shark swallows small fish for food. I started my own company 40 years ago. My best friend and I built a small painting factory literally in the ashes of World War II, and we worked there day and night. Gradually the company grew, we hired more staff, orders expanded, and so did our lines of business. By the 1960s, when Japan's electronics revolution was ready to take off, my company was big enough to become a major parts supplier to one of the nation's largest and best known electronics companies. What I didn't realize was that as soon as I succeeded in becoming such a supplier, I was considered part its "family." I was expected to be loyal to that company no matter what the sacrifice. For example, I was forbidden to accept orders from any other company, even when times were slow at my main client and my equipment sat idle.

I went to the president of this big company, not to demand my freedom but to profess my sincere desire to support his company's growth and to ask for more work. "Your words are like an expression of affection from an ugly woman," he answered; my willingness to serve his company was of no concern. My loyalty was taken for granted.

I was furious. I tried setting up an independent company to do work for other clients, but my "master" company sent managers around to threaten me: "No more orders, no more money, and a bad reputation throughout the business. Is that what you want, Sakai?" I must have been young and stubborn, for I stood firm against the threats of this big keiretsu, something few Japanese companies could ever contemplate.

My old partner (who was by now running our second company) backed me completely, and we decided to keep our businesses independent as far as possible. Needless to say, we saw some hard times after that, but we also grew, and as we grew, I learned the secret of good management that the keiretsu will never understand: the power of smallness.

Many years ago, I explained this simple reality to a visitor from the United States. "How can it be?" he asked. "Why doesn't the subcontractor just refuse to cut its prices and instead sell to another large company?"

This response, unthinkable in Japan but natural for an American, left me speechless. It made me realize how great a gulf lies between our two systems–and how little most people understand about Japan.

Han and Keiretsu

What we call "modern" Japan was born only about 100 years ago. For centuries before that, the nation was divided into small feudal fiefdoms called *han*. Each han was under the control of one man–the *daimyo*–who lived in a castle town surrounded by the agricultural land that provided his tax base and his power. The daimyo's extended family included people who were related only by marriage and even some who were adopted–but all were entitled to wear the family crest. Below the daimyo's family were his most trusted retainers–samurai of the highest rank who served the daimyo's household–then below them a lower level of samurai and, perhaps, another level.

On the bottom of this social pyramid were the hierarchies of the common people: farmers, artisans, merchants. The commoners, of course, outnumbered the daimyo's inner circle, and it was their work that provided the base on which he could build his castle and rule his fiefdom. Yet the commoners were treated more as property than as human beings. They were expendable resources to support the daimyo and his family.

The same kind of thinking that once dominated the han is at work today. The parent company in a manufacturing group thinks of itself as a daimyo, the supreme power, the apex of a pyramid in which production flows from the bottom upward and rewards from the top downward. The manufacturing daimyo's "family" of dozens of related companies includes some that are large and powerful in their own right–look at the Matsushita subsidiaries, many of which are listed on the stock exchange. Some family members may be related by corporate marriage or adoption.

Beneath the parent company's direct family are the trusted retainers–primary subcontractors–beneath which lies yet another layer of subcontractors. And beneath the subcontractors are layer upon layer of "commoner" companies, which have just a handful of employees whose only function is to produce a small quantity of goods–such as electrical parts–for the company just above them in the pyramid. These companies are tiny; their names and even their existence are unknown to the daimyo at the top. All it wants to see is the steady flow of production.

There was very little interaction among the han of old. Samurai were expected to show unwavering loyalty to their daimyo; they would not have dreamed of deserting their master even if there were work to be found elsewhere. The common people were also bound to their han; most were not even permitted to travel to other fiefs.

The modern corporate han are no different: vertical hierarchy is the hard and fast rule. Smaller companies produce for the next level up. No matter how bad times may get, companies can never leave their industrial group to seek employment elsewhere. And if they tried, no one would hire them–nobody likes deserters, even when they come cheap.

In short, today there is a Toshiba han, an NEC han, a Hitachi han, a Matsushita han, as well as a Toyota han and a Nissan han–a han for every major company in every major industry.

You can see why the Western idea of a small subcontractor trading with a large company on anything like equal footing is inconceivable in Japan. There are obvious benefits to the parent company from presiding over this pyramid and controlling suppliers' production levels and unit costs–which is why the industrial combines so jealously guard this system.

This historical metaphor of the han applies at a larger level as well. Just as groups of daimyo banded together under strong leaders to combine the power of their different han, all the giant manufacturers are members of larger industrial groups designed to further their common competitive goals. Some of these groups are descended from the old *zaibatsu*, the huge industrial groups that were disbanded after World War II. The names of their member companies are easily recognizable – the Mitsubishi and Sumitomo groups, for example. But outsiders have a harder time identifying other giant groups – the Fuyo Group companies, for instance – because they do not share a common name. In almost every case, keiretsu center around a main bank, which provides financing, information, and organization for the group.

While the old zaibatsu, with their enormously rich and powerful leaders, are dead, the keiretsu influence a wide spectrum of industries. Moreover, each keiretsu has its own distinct personality; to those of us who know them well, each one has certain recognizable features. Some keiretsu are very tight-knit, almost like an exclusive club. Their executives learn to speak a kind of group code, using words in special ways that only other members of the group will understand. If overheard in a restaurant, for instance, their conversation might sound odd or meaningless to keiretsu outsiders.

> When the guests at our party saw Asahi beer, they assumed my company was part of the Sumitomo group.

This group image can be so pervasive that it colors even the most ordinary business transaction. For example, one of my companies in a northern prefecture decided to throw a get-acquainted party in a hotel and invited representatives from dozens of local businesses. It was clear that the party was given to promote better relations with neighboring subcontractors.

We had a fine turnout, and everyone was cheerful – until one of the guests noticed bottles of Asahi beer, and only Asahi beer, on the tables. Most hotels and restaurants in Japan serve only one kind of beer because they buy from one supplier, such as Kirin, Asahi, Sapporo, or Suntory. But even the beer companies are members of giant keiretsu. So the large urban hotels are always careful to serve all four beers at such gatherings to prevent any hint of favoritism. Of course, at a small hotel in the distant provinces, having four separate suppliers just for beer would be a ridiculous extravagance.

Now, Asahi beer just happens to be a member of the Sumitomo group – the president is a former Sumitomo director. The sight of all those bottles of Asahi on the tables sent shivers through some of our guests. The general reaction was, "Why didn't you tell us this was a Sumitomo affair?" The whole mood of the gathering suddenly changed. My people tried to assure our guests that our company had no connection with Sumitomo, but the damage had been done – by the label on a beer bottle.

There are many ways in which the keiretsu affiliation is shown. Beer is just one example. The entire business world in Japan is separated, like lines on a battlefield, to indicate who are friends and who are enemies. If you belong to the Mitsubishi group, for instance, you not only drink Kirin beer to support a member of the group but you also bank with Mitsubishi Bank, buy securities from Nikko and life insurance from Meiji Mutual Life, drive a Mistubishi car, and insure it through Tokio Marine & Fire.

Sometimes this "keep business within the group" mentality becomes ludicrous. For example, Sumitomo Group members are very strongly encouraged to buy their home appliances from group member NEC. If they didn't, NEC (basically a computer and telecom maker) would be hard put to sell any refrigerators or TVs. And this goes right on down through the layers of subcontractors too. I'm sure there are many divisions of big companies that would go out of business if they didn't have thousands of captive customers forced to buy their products.

> Keiretsu are the ultimate force in Japanese industry.

Keiretsu are the ultimate force in Japanese industry. The manufacturing companies within them are controlled to some extent by the needs and policies of the group – just as each of the manufacturers has its own huge pyramid of subcontractors that functions as a group.

Cracks and Tunnels

All this strength among the industrial combines notwithstanding, strange things have been happening in Japan recently: the walls of the corporate han are still there, but cracks are starting to appear, and even the lines between the keiretsu are no longer inviolable. In large part, this stems from pressures put on the Japanese economy by the sudden doubling of the yen's value between 1985 and 1987. Because most Japanese manufacturers relied on exports for a large part of their income, they saw the sudden yen appreciation as a potentially crippling blow. Profits evaporated as the yen continued to climb. In order to recover their costs, they would have to double their export prices, which in turn would boost retail prices overseas, and ultimately the result would be a plunge in market share.

What did happen? Did retail prices of Japanese exports double?

No.

Did Japanese companies lose market share?

Not significantly.

Did the Japanese trade surplus disappear?

Not by a long shot.

So where did all the money go? What happened to those billions

Subcontracting 101

How does subcontracting work on a practical level? Consider a product like a personal computer, which requires scores of parts, high-level technology, and critical assembly. Even the most famous, top-selling PCs in Japan are made by subcontractors. How?

The parent company begins with feedback from its retail outlets to provide the basis for internal discussion of what kind of model to build next year. Say it decides it wants to make a 32-bit computer with this kind of drive, that much RAM memory, and so on. It draws up its own plans, its engineers make up blueprints, while other divisions decide on the quantity to be produced, the retail price, and thus the budget for producing it.

Then it gives parts of the blueprints to each of several small subcontractors. The biggest parts, of course, go to its immediate keiretsu partners, who then subcontract the components of those parts to smaller companies, who subcontract even smaller parts to other companies. Essentially, the machine is broken down piece by piece so that many small companies work on it. Almost no small company knows what it's making–just that the shipment is due on this date and for that price.

Even the first test models of the finished computer are made at the subcontractor stage because this is the level at which all the separate components are received from the smallest companies, then assembled and tested. Only at one of the larger subcontractors do the engineers know they're making a 32-bit PC with so much memory, a certain size hard drive, and so on.

One of the interesting byproducts of this convoluted manufacturing process is that it is common to make only a handful of test models of a new product. IBM, which controls its production much more closely, can make 100 or 150 test units and run them through every kind of test before manufacturing begins.

An NEC, Fujitsu, or Toshiba might make as few as 5 or 6 trial units because of the involvement of so many subcontractors in the premanufacturing system. Once their engineers are satisfied with the results, they can then set up the kind of mass-production assembly lines they're famous for. But even that will likely take place at one of their primary subcontractors.

upon billions of dollars–the difference between the Japanese exporters' plummeting sales revenue abroad and their costs at home?

The answer is clear to anyone who ever studied Japanese business seriously. The big companies pretended to tighten their belts. Many put out idiotic press releases about how they were saving scraps of note paper or using pencils down to their stubs to save money. But it was all a sham.

The real savings were achieved the old-fashioned way: the companies ordered their smaller subcontractors to cut prices. As the yen appreciated, they simply ordered more price cuts–and then more again. Yes, a lot of small companies went out of business. But what is that compared to the preservation of the Japanese economy? The large companies are the backbone of industry, or so the large companies insist. They must survive at all costs, they say. And survive they did.

The yen appreciation put a great deal of stress on the Japanese economy, and today deregulation in various sectors is keeping that pressure on. The result is an acceleration of the loosening of the keiretsu ties. The keiretsu are not gone by any means–but they no longer have absolute control as they once did. In hundreds of instances, in dozens of ways, the rigid walls between the keiretsu are fracturing, and individuals and companies (and business opportunities) are slipping through.

The biggest cracks in the keiretsu come from changes in workers' attitudes. For decades, Japanese workers have devoted themselves to one employer, working selflessly in the belief that big companies would always return their loyalty. When companies had to lay people off during the oil shocks of the 1970s, it became apparent that this was an illusion. Big companies take care of themselves first and their employees second. Young people today especially realize that big companies and impressive-sounding keiretsu are no longer a guarantee of anything. A majority of young people leaving college 15 or 20 years ago would be proud to join a prestigious group like Mitsui and wear a Mitsui pin in their lapels, regardless of there being better jobs elsewhere.

Today this "I'm a Mitsui man" way of thinking is disappearing fast. The university students who once cherished the "company man" illusion now think it's out of date. Gradually even the allegiance to the corporate group will fade. Already in the finance sector, for example, aggressive companies are beginning to hire talented managers from rival organizations, just as in the West, and some Japanese employees are forsaking company loyalty for a better job across the street.

In fact, the cracks in the keiretsu have been appearing for some time. More than 30 years ago, Matsushita Electric took advantage of the "company castle town" structure common to the factories of many big makers. Matsushita decided to "tunnel under" the walls of a powerful rival, Sumitomo. Matsushita was looking for a way to catch up with NEC–a key member of the Sumitomo Group – in the field of sophisticated communications technology.

Matsushita hit upon a bold and simple plan: it bought some land and built a factory near an NEC plant – producing the exact same kind of products as its established rival next door. It was a clear announcement to NEC's subcontractors that Matsushita was interested in the same kind of work and would welcome them as Matsushita subcontractors. The unwritten rules say that each major manufacturer is responsible for its own subcontractors and that manufacturers will not steal from one another. But Matsushita decided that if it continued to play by the old rules it would fall farther behind in an important technological field. By revising its corporate strategy, Matsushita succeeded brilliantly.

This kind of activity will only increase in the future. The keiretsu walls will never crumble. But more and more cracks will appear. And companies that are quick and aggressive enough to take advantage of the opportunities will grow and prosper.

A Chance for Foreign Companies

Cracks in keiretsu walls create opportunities for foreign companies – particularly U.S. businesses – and they should seize them. Japan is unquestionably the world leader in high-tech manufacturing. Japanese companies have skilled personnel and state-of-the-art equipment. But few foreign companies seriously consider manufacturing in Japan. Too many think only of the short term and the initial costs: real estate, construction, wages. They turn instead to Thailand, Hong Kong, Indonesia, Taiwan, or South Korea for their Asian production base. Many Japanese companies are making the same move – but they have no illusions about what they will find in these other countries: cheap mass production, not sophisticated high technology. They know that they will not find workers outside Japan who are anywhere near as well trained, well equipped, or reliable.

Forgive me for waving the Japanese flag for a moment, but when it comes to getting value for money in manufacturing, I'm an expert. I know exactly what I can expect from production in any one of a dozen countries, and I choose to keep my factories right here in Japan. I know that Japanese workers are uniformly diligent and honest. If you ask them to build something to exact specifications, finish it by a certain date, and deliver it on time, they'll do it. And they'll do it right.

> Most foreign companies are providing the technology and financing for their own rival – and paying twice.

Of course, some foreign executives stay away from Japan for a different reason than cost: they tell horror stories of having their technology "robbed" by a Japanese partner whom they trusted. When I hear these stories, I have to bite my tongue. I want to ask, "What did you expect?"

Too many U.S. and European companies license production of some important product to a large, well-respected Japanese company. The production is handled beautifully, and they have no complaints. But then, about two years later, the Japanese company comes out with its own version of the same product, becoming a competitor overnight. Then there are howls of outrage from abroad. Such an incident underlines the biggest problem with Western businesspeople: they don't learn the basics about how these large Japanese companies operate, so they find out the hard way when they try to do business with them.

For the record, here's what really happens behind the scenes. Let's say a U.S. company has something to produce – a piece of electronic machinery, for example. It licenses production to a big Japanese company, which then subcontracts the job to one of the group companies. By the time the original contract is finished, the subcontractor has a clone product ready to go, and the U.S. company will soon discover that it has a new rival.

What most foreign companies never realize is that they not only provided the technology and financed the creation of their rival but they also paid for it twice: the big company they gave it to probably collected a nice margin before handing the job over to one of its smaller group companies. This kind of thing has been going on for years.

Of course, it doesn't have to be this way. A foreign company could choose to bypass the big Japanese manufacturer, split up production among a couple of reliable smaller companies, and avoid the whole problem. The key to this approach is the basic interests of the small subcontractor: it doesn't want to compete with you – it wants more orders from you.

A really smart foreign company would go all the way: buy some land, put up a factory, and hire a work force right here in Japan. If a U.S. company put up capital to start operations in Japan, if it hired a few Japanese managers to run the local operation and had a few people around who could translate orders into Japanese, it would make the same profits as the big Japanese companies do – maybe even more. Would it make a difference to the workers that the company was American-owned or had an American executive from Tokyo to oversee it? Not at all. Like any other smaller business, the workers – and the managers – would simply want a chance to show what they could do without having someone always demanding they cut prices.

The Spartacus Strategy

Some years back, there was another popular movie called *Spartacus* – a story of a rebellion among the slaves in ancient Rome. From that story, let me offer a modest proposal, one I believe could actually achieve substantial results in a short period of time.

The U.S. government should establish an office in Tokyo with the single mission of identifying small and medium-size Japanese companies that are suitable for U.S. manufacturers to use as subcontractors. Let it be known that this office is hard at work at its mission and that its recommendations are going to the widest possible range of U.S. companies. Set up a special telephone number that Japanese manufacturers out in the farthest provinces can call to find out more about the program and to offer their services. Have Japanese and Japanese-speaking staff on hand to answer their questions and to reassure these local businesspeople that the program is legitimate and that their inquiries will be kept confidential. Many of the small subcontractors from the distant provinces have never seen a foreigner, much less talked with one. Let this agency act as a liaison between U.S. parent companies and Japanese manufacturers, helping to arrange introductions, provide language assistance, make sure specifications are clear to the subcontractors, and tackle other go-between work.

Remember, these rural manufacturers are intelligent and hardworking–but most have spent their entire lives inside a single corporate hierarchy. The idea of doing work for a foreign company unconnected to the Japanese system may seem strange to them at first. But they will soon see the benefit of breaking their dependency on a single "master" company. If Western companies are willing to show that they appreciate the quality of work done by the smaller "slave" subcontractors and that they will abide by agreed-on prices–unlike the Japanese "master" companies–word will quickly spread and more companies will come forward.

This will affect U.S. companies on two levels. On the micro level, the companies will profit directly. They will develop stable, reliable, innovative sources of supply for critical technologies. They will even acquire some surprising new technologies–because many of Japan's top subcontractors have developed innovative processes that the giant manufacturing companies have then stolen. If U.S. companies begin to do business with these subcontractors, some of these technologies can flow back to the United States. But perhaps most attractive of all, U.S. companies will discover that they can get all of these benefits at a fair price.

The way to create change is to get inside–and free Japan's indentured servants.

On the macro level, this opening up of opportunities for Japan's small and medium-size businesses will accelerate the erosion of the foundations of the keiretsu–particularly in the all-important manufacturing sector. No, the keiretsu will not come crashing down. But internal pressure for change will increase dramatically.

Foreign governments and foreign companies that want to change critical elements in the Japanese economy are misguided if they think threats from outside will do it. Outside threats only help to unify the powerful masters inside–and they just become more powerful.

The way to create change is to get inside. Once inside, it is possible to offer the indentured servants an opportunity to escape, to work as they choose for whomever they choose–and to unleash the pent-up power of Japan's small and medium-size companies. Doing that alone would be a big step toward loosening the grip of the keiretsu.

Reprint 90609

To stay competitive, stay home.

Manufacturing Offshore Is Bad Business

by Constantinos C. Markides
and Norman Berg

July 1985: AT&T decides to transfer production of residential telephones from its only U.S. telephone manufacturing plant, in Shreveport, Louisiana, to Singapore.

February 1986: United Technologies announces it will close its diesel-engine parts plant in Springfield, Massachusetts and transfer operations to a nonunion plant in South Carolina and two plants in Europe.

February 1987: General Motors plans to phase out the production of A-body cars in the United States and move it to its Ramos Arizpe, Coahuila plant in Mexico.

For decades, foreign direct investment has been a common practice among American companies, so investments like these don't seem particularly noteworthy. But they are. In the past, U.S. companies went abroad primarily to secure a foreign market or to obtain raw materials. Now they go overseas to buy or make products and components to ship back to the United States. The new investments are not complementing domestic production; they are replacing it.

Constantinos C. Markides is a doctoral candidate studying business policy at the Harvard Business School. His research centers on U.S. competitiveness in the world economy. Norman Berg is professor of business administration at the Harvard Business School, where he teaches in the Program for Management Development. His longtime interest is in conglomerate management, a subject on which he has frequently written.

Manufacturers defend sourcing from overseas as the only way to compete with inexpensive, high-quality imports. They say that moving to cheap-labor countries like Mexico, Taiwan, and Malaysia for export back to the United States is allowing U.S. industry to regain its world standing. Economists generally approve. They consider the migration to low-wage areas an adjustment caused by changes in international comparative advantage.

Companies should face manufacturing at home—like Kodak and Black & Decker.

Others are less enchanted by the trend. Labor unions claim that it deindustrializes the country and destroys American jobs. Some observers have argued that it "hollows" the nation's industrial base and threatens the standard of living.

In the debate over whether offshore manufacturing is good for the nation, it is always assumed that it is good for an individual company. But we challenge that notion. The mere fact that a lot of companies are doing it doesn't make it smart. Going overseas is hardly the panacea many people think it is. At best, it is just another quick fix. In their rush to save money, managers often lose sight of the high penalties of moving abroad. And by continually shifting manufacturing to the areas with the lowest labor costs, they are merely postponing the inevitable day of reckoning when they must confront the parts of the business that really need reform.

It's Not the Only Option

American manufacturers claim that going offshore is their only alternative if they are to stay competitive against foreign rivals. It's either offshore manufacturing or no manufacturing at all. As powerful as this argument may seem, it doesn't hold up well in light of several facts.

First, an increasing number of U.S. companies are responding to the import threat by improving their competitive position at home. They are convinced that creating long-term competitive advantage requires a commitment to a new way of doing business–not just a shortsighted attack on labor costs.

Take Eastman Kodak. After a disappointing 1985, in which earnings dropped 31% from the year before (excluding the $563 million Kodak lost when it withdrew from instant photography), the company decided to drastically change its strategy. While still relying on some offshore manufacturing, Kodak also embarked on an aggressive and multifaceted program to restore its competitive position at home. It made an all-out attempt to create long-term competitive advantage by addressing the business as a whole, not just isolated parts of it.

Consider some of the steps Kodak has taken or plans to take: reduce overhead by trimming employment, especially in middle management; revise the wage dividend plan; cut operating and expense budgets; eliminate inefficient operations and marginal product lines; reorganize internally; increase R&D expenditures; move into new technologies; introduce new products; and improve quality and cost efficiency. The effect of these changes was immediate: in 1986, sales grew by 9% to $11.5 billion, while earnings from operations climbed 24% to $724 million. Earnings per share were $3.52 in 1987, versus only $1.10 in 1986.

Black & Decker is another case in point. It reduced its work force by 40%, consolidated operations to achieve economies of scale, eliminated hundreds of administrative positions at corporate headquarters, modernized plants, introduced new manufacturing methods, standardized models, expanded and upgraded marketing and engineering capabilities, and moved aggressively into the low end of the professional power-tool market. The results: for fiscal year 1986 the company reported net earnings of $6.3 million–that compares with a net loss of $158.4 million the previous year. EPS went from 49¢ to 95¢ between 1986 and 1987.

Litton Industries is making similar changes in its operations to fight foreign competition in microwave ovens. The company has redesigned its product line, improved the quality of its products, cut down on labor costs, and introduced new models.

Some companies have embraced automation to stay competitive. In 1985, GM had more than 4,000 programmable robots in operation; it expects to have 10,000 in place by the end of 1988. Xerox has spent more than $100 million to automate manufacturing and materials handling, and Fairchild Semiconductor moved its assembly operations back to the United States after automating the welding of semiconductor chips and the inventory tracking system.

Second, more Japanese companies are building manufacturing plants in the United States at precisely the time when many American companies are claiming it is impossible for them to stay home and be competitive. Every major Japanese automaker has an assembly plant in the States. By one count, there are more than 600 Japanese plants operating on American soil. Honda's plant in Marysville, Ohio, Nissan's plant in Smyrna, Tennessee, and Mazda's plant in Flat Rock, Michigan are but a few. Japanese electronics companies are also operating in the United States. Fujitsu, Hitachi, Mitsubishi, NEC, Toshiba–all have assembly or manufacturing facilities in America.

> **If the Japanese can manufacture in the United States, so can Americans.**

The Japanese are investing in the United States for four reasons: to increase their political clout and prevent further trade restrictions by creating jobs for Americans; to ensure access to the American market in case exports to the United States are restricted further; to get a better feel for their most important export market so they can be more responsive to it; and to hedge against fluctuations in the value of the dollar.

To be sure, the Japanese in the United States enjoy some advantages over existing U.S. plants. Because of a younger work force, they have low pension expenses and low insurance costs. They also have newer facilities. But these benefits are not the chief reason Japanese factories are competitive. The plants are successful because of their manufacturing techniques. Studies have shown that the main reason the

The Race to Manufacture Offshore

It's a fact that the electronics, textile, machine tool, subcompact auto, and other industries are moving their manufacturing operations abroad. The evidence is all around us. Although no one set of statistics accounts for all the forms of offshore manufacturing–be it setting up plants abroad, purchasing from a foreign subsidiary or joint venture, subcontracting from a foreign company, or buying from foreign companies through market transactions–the available data are persuasive.

The best information on products manufactured or assembled abroad is a set of numbers from the U.S. International Trade Commission on imports entering under tariff items 806.30 and 807.00. These provisions permit the portion of the product made of U.S. components to enter the United States duty free. A quick glance at the 806 and 807 statistics shows a dramatic rise in the value of offshore-manufactured imports (the accompanying table of selected years illustrates the trend).

Offshore imports rose continually from $953 million in 1966 to $36.5 billion in 1986, despite fluctuations in the dollar. Their share of U.S.-manufactured imports nearly doubled in the same period.

Most of these imports are in three price-sensitive industries: autos (59% of the total in 1985), electronics (15%), and textiles (3%). Most of the auto industry's investments have been in Europe, Japan, Canada, and Mexico. The electronics industry has invested mostly in East Asia, while the textile investments are concentrated in Latin America and the Caribbean.

These figures grossly underestimate the true extent of the phenomenon. They exclude, for example, products of a U.S. company manufactured abroad containing only foreign components. Such items are considered ordinary imports and don't fall under the 806 and 807 provisions. Similarly, products of a foreign company under contract with a U.S. business enter as ordinary imports. And much of what a U.S. affiliate abroad ships to its domestic parent does not appear in the 806 and 807 statistics. (Note that at least 35% of U.S. imports and exports from 1977 to 1980 were intracompany.*)

Obviously, U.S. companies are augmenting their overseas facilities: capital expenditures by majority-owned foreign affiliates of U.S. manufacturing companies climbed from $12 billion in 1978 to $17 billion in 1986. Meanwhile, U.S. manufacturing capacity utilization fell from 84.2% to 78.8%.

*Jane Sneddon Little, "Intrafirm Trade and U.S. Protectionism: Thoughts Based on a Small Survey," *New England Economic Review*, January-February 1986, p.42.

Aggregate statistics aside, evidence abounds. In 1974, some 70,000 workers were employed in 450 plants along the Mexican border under the in-bond, or *maquiladora*, program. By 1986, the figures had grown to 300,000 workers and 1,100 factories. The value added by these plants grew from $300 million to more than $1.3 billion in those 12 years. American businesses, operating 865 of these factories, account for most of the production.

In 1985, U.S. companies like RCA, Motorola, and Texas Instruments employed more than half the 73,000 people working for the electronics industry in Malaysia and sent back to America more than $300 million worth of semiconductors. In Mexico, the three U.S. automakers accounted for more than 55% of Mexico's total car production in 1986, and all three were planning to expand there. U.S. companies' imports from South Korea, Taiwan, Mexico, and Brazil are estimated to rise from 100,000 units in 1984 to 500,000 in 1988.

The Growth of 806 and 807 Imports

Year	*Total Value of 806 and 807 Imports (in millions of dollars)*	*Percentage of Total U.S. Imports*	*Percentage of U.S.-Manufactured Imports*
1966	$ 953.0	3.7%	6.4%
1967	1,035.1	3.8	6.5
1969	1,838.8	5.1	8.0
1970	2,208.2	5.5	8.5
1973	4,247.1	6.0	9.4
1975	5,162.4	5.3	10.1
1978	9,735.3	5.5	9.1
1982	18,275.5	7.4	12.1
1983	21,845.7	8.1	12.8
1985	30,535.1	9.0	12.3
1986	36,469.9	9.9	12.4
1987	39,820.1†	10.0	12.9

Sources: U.S. Tariff Commission, *Economic Factors Affecting the Use of Items 807.00 and 806.30 of the Tariff Schedules of the United States,* Publication 339 (Washington, D.C.: USTC, September 1970); U.S. International Trade Commission, *Imports Under Items 806.30 and 807.00 of the Tariff Schedules of the United States, 1982-85,* Publication 1920, and *The Use and Economic Impact of TSUS Items 806.30 and 807.00,* Publication 2053 (Washington, D.C.: USITC, December 1986 and January 1988); and U.S. Department of Commerce, *Survey of Current Business,* various issues.

†Projected figure.

Japanese were able to dominate the market for small cars over the last decade was not because of higher capital investment rates or more advanced technology but because of management philosophy and excellence in manufacturing.[1] By emphasizing superior product designs, high quality, minimum inventories, waste elimination, and worker participation, the Japanese emerged as the cost and quality leaders in the industry. Now, using the same techniques, the Japanese are outperforming domestic rivals on their home turf.

The ability of the Japanese to manufacture in the United States raises serious questions about the rationale American companies use for going offshore. In particular, if the Japanese can manufacture in the United States and still be competitive, why do U.S. companies have to go offshore to stay alive? More important, if U.S. companies have the option of becoming more competitive by improving manufacturing at home, why do they choose the quick fix of manufacturing offshore?

Third and last, Japanese companies have managed to stay competitive over the years without resorting to offshore manufacturing. Overseas production of the Japanese semiconductor industry, for instance, has amounted to less than 4% of total production.[2] Even when the yen got strong, the Japanese found it unnecessary to search for low-cost labor. They stayed world-class competitors by investing in automatic wire-bonding machines and by using flexible manufacturing.

In consumer electronics, the Japanese didn't go offshore but instead used intelligent product design and simple automation to stay competitive. In the early 1970s, while U.S. color TV manufacturers like RCA, GE, and Zenith were rushing to the Far East, Japanese manufacturers switched to 100% solid-state chassis, automated insertion and testing, and reduced component counts through extensive use of integrated circuits, early use of in-line tubes, and single-circuit board designs.

> Lowering wages helps a little – but labor is only 15% of total costs.

The few Japanese companies that chose to go offshore did so either to supply their export markets or, more commonly, to get around trade restrictions. Japanese textile companies went to Asia in the 1970s, for example, to avoid OMAs (orderly marketing agreements) imposed on exports of apparel to the United States and Europe. Similarly, Sanyo, Sony, and Hitachi have been moving manufacturing and assembly operations to Mexico to circumvent U.S. trade laws and to preempt American protectionism.

In the period 1985 to 1987, when the yen gained by more than 60% against the dollar, 53% of the 214 Japanese manufacturers responding to a Keidanren survey stated that they would cope with the strong yen by shifting their emphasis to domestic markets. They planned to do so by upgrading their products, incorporating greater value added, and minimizing costs. More important, 73% of the companies surveyed reported that they will not resort to more outside contracting, and 77% reported that they do not plan to cut wages. The 29% that planned to shift some production overseas were doing so to supply export markets, not the Japanese market. Nearly three-fourths of the companies expected to replace or had already replaced outside subcontracting with internal production. This behavior is in sharp contrast to that of American companies in the early 1980s when the dollar was strong.

The Bush Is Full of Thorns

Offshore manufacturing is not, then, the only option available to companies under competitive siege. It is, in fact, a poor option for many organizations. Managers should know that offshore manufacturing is not all roses and labor savings. Before they make the move, they should have a fuller picture of where they're headed. They should know that there are some powerful reasons *not* to go offshore.

The savings can cost a lot. Granted, companies can often save money on labor and materials by purchasing or manufacturing overseas, but other costs – some not so obvious – may well offset the gains. Offshore sourcing usually involves larger inventories, for example, and higher administrative costs. Parts made overseas are less likely to meet specifications, so quality costs may be higher. Factor in higher transportation expenses and tariffs, and don't forget the cost of training foreign workers.

It takes longer to get supplies from an offshore location, so companies operating abroad are slower to respond to changing market demands. That too has its price. In 1984, some big U.S. retailers were stuck with huge inventories of imported goods they had

1. See, for example, William J. Abernathy, Kim B. Clark, and Alan M. Kantrow, "The New Industrial Competition," HBR September-October 1981, p. 68.

2. Dennis J. Encarnation, "Cross-Investment: A Second Front of Economic Rivalry," *California Management Review*, Winter 1987, p. 38.

3. Kenichi Ohmae, *Triad Power* (New York: Free Press, 1985), p. 5.

ordered the year before—when a slowdown in consumer spending was nowhere in sight!

These costs may seem obvious, but companies rarely take them into account when making the offshore manufacturing decision. According to a survey commissioned by the National Tooling and Machining Association, American tool and die makers routinely ignore these "hidden" costs of offshore manufacturing. Yet according to the survey, these costs typically add 5% to 15% to the foreign vendors' bid price for shipping, 3% for additional paperwork and communications, 5% to 10% for added inventory, and up to 35% for unanticipated design changes.

Other costs are less direct. Shifting some production operations overseas may prevent the company from exploiting economies of scale at home as well as abroad. The implications can be especially grave during a recession. Going offshore can also cause underutilization of existing manufacturing capacity and, ultimately, plant closings and layoffs.

Finally, companies that move outside the United States may lose valuable customers, who may switch to other U.S.-based suppliers. Italian chip maker Dynamit Nobel Silicon is among the foreign companies that opened a factory in the United States to be closer to its American customers. The managing director explained the decision by saying that "the distance between Italy and California is such that [customers] do not want to get any more than 10% of their needs from us."

You don't really save much on labor. On the surface, looking for low-wage sites for manufacturing is a logical way to reduce total production costs. But workers in less developed countries tend to be less productive than Americans, so a straight comparison of wages gives an inaccurate reading of the potential savings. Worse, in most businesses, direct labor is no longer a significant portion of total costs.

A survey of manufacturers by the National Association of Accountants found that on average labor represents only 15% of the cost of making a product. For most electronic items, labor is only 5% to 10% of the total cost. The wage savings are therefore unlikely to have a big impact overall. In some cases, they are offset by higher transportation costs alone. According to one expert, the typical cost of transporting a color TV set from Southeast Asia to the West Coast is 13% of its value at the time of shipping; making the set in Asia saves just 10%.[3]

Managers' preoccupation with labor costs deflects attention from the other 85% of the cost structure. Opportunities to save money in administration, inventory control, marketing, R&D, and distribution far exceed those relating to labor alone, but they are often overlooked.

You'll hollow the corporation. The semiconductor industry has split into two market segments: commodity and semicustom. Advocates of offshore manufacturing claim it doesn't matter if the commodity side of the business goes offshore in search of lower wages. As long as the design and R&D talent remains in the United States, so will the value added. Other industries have adopted this argument as well: as long as the company focuses on innovation, advanced technology, and excellent service, it doesn't really matter if it manufactures products abroad.

But a business cannot design in a vacuum. It cannot exploit new technologies if it has no chance to apply them. And it needs the profits from commodity production to fund R&D. The fact is, design and manufacturing are linked. A company that subcontracts its manufacturing to foreigners will soon lose the expertise to design and the ability to innovate, because it won't get the feedback it needs. Moving engineering offshore along with manufacturing is not a solution; it just accelerates the process. When companies do that, they give potential competitors not

"Looks like the gods are angry."

only finances and managerial expertise but also engineering skills.

The TV industry's inability to design and develop the new generation of TV products–the videocassette recorders and camcorders–is the perfect example. The U.S. television industry began assembling black and white sets overseas in the late 1960s. Assembly of color TVs soon followed black and white, and manufacturing followed assembly. By 1987, not one U.S. company was producing black and white TVs domestically, and only two–Zenith and Curtis Mathes–were making color TVs. Many people contend that the move offshore dispossessed U.S. manufacturers of manufacturing and design technologies needed to innovate or even compete with the innovators.

The semiconductor industry has fallen into the same trap. Americans have all but given up on the production of high-volume semiconductors such as 64K RAM chips. But memory chips are the cornerstone of semiconductor technology. Because they are a high-volume product, they serve as a testing ground for engineers trying to produce new technologies for other applications and for manufacturers trying to perfect delicate processes. In addition, they generate earnings for further research. By abandoning the commodity memory products, electronics companies put their design capabilities and technological leadership at risk.

Even normal product development suffers. As a senior executive of a large corporation explained: "I don't think people realize when they make the offshore decision that it is really a commitment to freeze the product. There is no way to make rapid design changes and product updates at a remote location."

Meanwhile, collaborators become competitors. The same senior executive continued: "To survive, the offshore manufacturer must build his own design or technology capability, and very quickly the game is over. He has the capability and the market." Consider Taiwan's Sunrise Plywood and Furniture. For years the company acted as an export platform for California's Mission Furniture. It relied on Mission's designs and blueprints to manufacture furniture suitable to American tastes. Now the Taiwanese company is one of Mission's competitors, exporting directly to the United States through its own marketing subsidiary.

The pattern is widespread. Hitachi, which has made microprocessors under license from Motorola, is now introducing its own 32-bit microprocessor. Toshiba, which acted as a supplier of copying machines to 3M, is now promoting its own brand name. Singatronics, which for years produced electronic games and pocket calculators for multinationals, is now pushing its own proprietary line of electronic medical instruments. And Daewoo, while still a subcontractor to U.S. companies, now sells its own personal computer.

The danger that collaboration will give way to competition is immediate and real. As the newly industrializing countries of Asia lose their advantage as low-wage producers to places like China and Thailand, they are increasingly anxious to develop their own technology-intensive industries and marketing capabilities. One executive of a large multinational told us, "Many Americans are naive about how insistent other nations have become on developing a full capability. Even in aerospace, it is hard to satisfy a coproduction requirement anymore by just letting the foreign plant rivet an aluminum assembly. The host country wants the whole technology, and it is not long before a competitor has been developed."

Many offer U.S. manufacturers incentives to bring their technology with them. Taiwan offers R&D facilities plus attractive loan packages. Malaysia has stepped up its efforts to get companies to invest in research, and Ireland has been encouraging foreign businesses to boost their product development efforts there.

U.S. companies that transfer technology across national borders don't have the same protection against piracy they enjoy at home. Indonesia, for example, has no patent protection at all. Korea denies copyright protection to software, semiconductors, or foreign works. Other countries require an importer of technology to license local companies to use that same technology for modest fees.

> First you build an offshore plant, then you give away your technology–then you're out of business.

The advantage doesn't last. Offshore manufacturing is most promising when three conditions hold: the dollar is strong, foreign wages are low, and trade barriers are absent. None of these factors is within a company's control. Most of the offshore investments between 1982 and 1985 were motivated by the strong dollar. Now that the dollar is considerably weaker, running those operations is more expensive. U.S. companies that purchase or produce abroad are not as price competitive as they were three years ago; the price of their products has gone up, along with those of all the other imports. There is no simple solution to their dilemma. Deere & Company, for example, makes its small tractors for the U.S. market in Japan and its midsize models in Germany. As the dollar declines, Deere will have to either raise prices or lower profit margins. Indeed, the aggregate figures for Deere's return on sales show a marked drop between 1984 and 1986: from 2.3% to −6.3%.

Reducing offshore activities when the dollar weakens is not a realistic option. Most of the offshore investments are irreversible capital expenditures, and they are hard to liquidate. But even purchasing arrangements are hard to change overnight. And it's unwise to sever a sound relationship with a foreign supplier when you know that the dollar may strengthen in a year or two.

Moreover, foreign wages inevitably rise. As workers learn and become more productive, they command higher pay. And as foreign countries grow economically, workers want a bigger piece of the pie. In Mexico, the minimum wage was increased three times in 1986–in January, June, and October. In Taiwan, average monthly earnings in manufacturing quadrupled in a ten-year period, from 2,929 Taiwanese dollars in 1974 to 12,844 Taiwanese dollars in 1984. In Korea, employee compensation doubled from 1979 to 1984. U.S. companies could run themselves ragged chasing low wages from one country to another.

Remember too that the threat of protectionism always looms. Importers are always risking a toughening of import restrictions. When a country's trade surplus with the United States swells, protectionism pressures intensify. Taiwan's $15.7 billion surplus in 1986, for instance, heightened pressures in the United States to impose trade restrictions on Taiwanese products–even though most of these products are exported to America by U.S. companies, including GE (Taiwan's biggest exporter), IBM, Hewlett-Packard, and Mattel. Such trade restrictions could have wiped out any cost savings from subcontracting or operating in Taiwan. To avoid rough treatment from the U.S. administration, Taiwan eventually increased the value of its currency, a move that made offshore imports originating in Taiwan more expensive.

U.S. companies could become gypsies, moving from one location to another as the cost of protectionism rises. U.S. retailers and clothing importers, for example, have embarked on systematic island-hopping, moving from one island to another to bypass limits on clothing imports. As soon as a new source of merchandise is found, the U.S. government moves in and imposes quotas, forcing the American importers to move on to other islands. This could be an acceptable short-term strategy, but sooner or later you run out of islands.

You may get trapped. Once a company moves its manufacturing operations to a developing country, the host government may begin to pressure management to transfer more advanced technologies or to support local spin-off industries. In many instances, the host countries insist on domestic content, technology transfer, and domestic equity positions that eventually lead to independent production capabilities. The company is often trapped: if it wants to stay in that country, it has no choice but to accede.

> Overseas cost advantages are fleeting: wages rise, and the dollar weakens.

Mexico is a case in point. In early 1985, it rejected IBM's plan to build a plant there to produce microcomputers. But when IBM agreed to increase its plant investment from $6.1 million to $91 million, to buy parts built in Mexico, and to export 92% of its production, the government reversed itself. In the auto industry, Mexico has a 60% domestic content requirement for cars produced for the domestic market and a 30% domestic content requirement for cars destined for export. In addition, foreign businesses are limited to 40% ownership of joint ventures in auto parts.

Consider also India. In 1973, India passed a new law, the Foreign Exchange Regulation Act, under which foreign companies operating in India had to dilute their equity positions. When in 1977 the government asked IBM to dilute its equity to 74% and Coca-Cola to 40%, both companies decided to leave India rather than comply. Coca-Cola didn't want to disclose its syrup formula, and IBM didn't want to lose control of its marketing operations. India didn't miss the two companies: Burroughs and ICL are now doing IBM's work, and an Indian soft drink company called "77" has taken over Coke's market. But IBM and Coke no doubt miss India.

You'll lose valuable friends at home. When companies move operations overseas, life at home changes. Labor unions, resenting the loss of domestic jobs, are less likely to cooperate on other fronts. Their dissatisfaction is particularly important because U.S. companies need their help if they are to become more competitive. As UAW President Owen Bieber put it in a 1986 speech, "If we can't get support from the main players in the industry on our public policy agenda, the question might be asked why we should continue to work with them on the productivity side."

Labor dissatisfaction can also be expensive. In November 1986, the UAW struck GM's Delco Electronics plant in Kokomo, Indiana over a plan to boost outsourcing from Mexico. After a seven-day strike, GM canceled the plans. The National Association of Machinists and Aerospace Workers has proposed a "Rebuilding America" program that would force companies to contribute 1% of after-tax profits to a fund used to create new industries to replace those that leave the country. Similarly, the Teamsters union has proposed a plan that would require U.S. companies making products abroad for import back to the States to pay their foreign workers at least the going U.S. wage or forfeit the privilege of access to the U.S. marketplace. American companies paying less than the prevailing U.S. hourly compensation would have to put the difference into a fund for retraining U.S. workers.

Clothing importers have been island-hopping across Asia – forced to move each time wages rise.

Pressure can come from society as a whole. As more companies go offshore, more jobs are lost and the trade deficit worsens. Growing unhappiness over these conditions can create pressure on the government to intervene and force companies to assume part of the social costs they are generating. The government might, for instance, pass legislation forcing companies to give their workers "early" notification of plant closings or to assume responsibility for the training of their laid-off workers.

Offshore manufacturing has another effect of political significance: companies that may have lobbied collectively five years ago may become divided on issues of protectionism. That is, those who import most of their products and components have different concerns from their industry counterparts that manufacture in the United States. The whole industry loses clout. The semiconductor industry experienced such a split. In June 1985, Micron Technology Inc. of Idaho filed an antidumping petition against seven Japanese producers of 64K DRAM chips. The Semiconductor Industry Association, however, did not support the move because some of its members – like Motorola and Texas Instruments – produce in Japan. The association's official policy is to improve member access to the Japanese market.

The auto industry is another divided industry. GM, the industry's leader in forging international alliances, has vigorously opposed protectionism. Ford and Chrysler, on the other hand, have repeatedly asked for surcharges on Japanese imports. Similarly, in the textile industry, apparel manufacturers and big retail stores have paid no attention to the "Crafted with Pride in U.S.A." campaign that the textile industry and labor unions are promoting. Why should they, when 20% of their brand-name products are imported?

Of course, offshore manufacturing isn't wrong in every case. Indeed, there are legitimate reasons to locate overseas – to take advantage of certain natural resources, to expand export markets, or to be more responsive to local markets. But managers should question the assumption that it is always right. Chances are, in their eagerness to save money on labor, many companies are giving up more than they bargained for, not the least of which is their future competitive position.

Companies must recognize that offshore manufacturing does not constitute a long-term strategy. At best, it is merely a short-term tactical move that buys time for companies to restore their competitive health at home. Unless they break away from old traditions and look at their business as a total package, American companies cannot expect to become world competitors.

Reprint 88510

Stories from the Front Line

Investments in technology and people can keep jobs and profits in the United States.

Cold Competition: GE Wages the Refrigerator War

by Ira C. Magaziner and Mark Patinkin

Fifty miles south of Nashville, outside the city of Columbia, where the restaurants offer Bar-B-Q and catfish, is an unlikely piece of smokestack America. There, nestled amid the pine and hardwood of rural Tennessee, is one of the world's most automated factories. Had it not been built, U.S. households might soon have had yet another product–the refrigerator–stamped "Made in Japan." Instead, here in the heartland, General Electric found a way to build products better and cheaper than those made by foreign workers paid one-tenth American wages. The going has not been easy, but GE's struggle shows the challenges the United States must and can meet if it is to regain world manufacturing leadership.

Tom Blunt still remembers the day in 1979 that he first stepped into Building 4, the plant in Louisville, Kentucky where compressors for GE's refrigerators were made. The compressor–the pump that creates cold air–is by far the most expensive part of the product. It is also the refrigerator's heart, as important as an engine in a car. You'd have never guessed that by looking at Building 4.

The plant was a loud, dirty operation built with 1950s technology: old grinders, old furnaces, too many people. Finishing a single piston took 220 steps. Even the simplest functions had to be done by hand. Workers loaded machines, unloaded machines, carried parts from one machine to the next. The scrap

DRAWINGS BY MARK STEELE

rate was ten times higher than it should have been; 30% of everything the plant made was thrown out. There was only one thing Blunt liked about Building 4. He liked the thought of beating it–changing it, rebuilding it. But it wasn't his place to suggest that. He had only recently joined GE's Major Appliance Business Group (MABG) as chief manufacturing engineer for ranges. He was too new to start pushing for major projects–especially projects in someone else's department. Besides, he thought management would never pour huge dollars into redoing a whole factory. He'd come to realize that MABG preferred cosmetics–bells and whistles–to engineering. In Louisville, the money was all on the marketing side.

> Ten years before, sourcing would have been a sacrilege. Now it looked like salvation.

Over the next 18 months, however, others at GE also began to worry about Building 4. A series of warning signals started to wake Louisville up. The group's profits were off. Market share was falling. Competitors were pushing hard on several fronts. Matsushita was manufacturing better, cheaper compressors in Singapore–and selling them to GE's Canadian subsidiary. Mitsubishi was experimenting with rotary compressors–a technology that GE had invented but used only in its air conditioners. Most disturbing, Whirlpool, GE's chief competitor, was moving its compressor manufacturing to Brazil. While Louisville had been focusing on bells and whistles, Whirlpool had looked overseas, glimpsed the future, and acted.

Then, the threat got even closer to home. In the fall of 1981, both Matsushita and Necchi, an Italian manufacturer, approached MABG itself, offering lower priced compressors that were indeed fine machines. Had that happened ten years earlier, there would have been one response: fight back. No one would have thought of buying from the competition. Even to whisper the word "sourcing" would have been sacrilege. MABG manufactured at home–in the United States. Its factories were unequaled. But things had changed. Now many in Louisville began to wonder if Japan could be their deliverance. People began talking about a new strategy–sourcing.

Tom Blunt didn't appreciate that kind of talk. By now he'd been named head of advanced manufacturing for refrigerators, MABG's biggest product. Few things put him in a worse mood than a decision to close a plant. "Sourcing makes sense in some circumstances," he says, "but you can't source everything. My instinct is always–always–to *make* things." Colleagues told him it was time to face the truth: there are certain areas, certain products that the United States can't compete in anymore. "Bull," he would say. "All we have to do is find a way to make it faster, cheaper, better."

John Truscott, MABG's chief engineer, agreed. "The compressor is the heart of the refrigerator," he says. "The refrigerator is the heart of this group. I didn't want to give away our heart." Actually, no one was quite ready to go that far yet. Sourcing was a compelling idea, but it was still a new one. MABG needed more information. My consulting firm was asked to supply it.

My research began in Japan. I found that Matsushita had built a plant in Singapore that planned to produce millions of low-cost reciprocating compressors for world markets. Both Toshiba and Mitsubishi were making rotary compressors that were cheaper, quieter, and more efficient than the reciprocating compressors all other refrigerator manufacturers used. Sanyo was planning to move into rotaries as well. None of this was a fluke. Obsessive exporters, the Japanese had spent the last few years traveling to the United States, probing the appliance market in search of a weakness–and found it. Now they were anxious to show off their plants. They knew that GE was considering sourcing compressors, and they all wanted to sell.

Next I went to Italy; Necchi turned out to be as great a threat as the Japanese manufacturers. Its new compressor plant was far more automated than Building 4. Finally, I visited Embraco, the new Whirlpool plant in Brazil. GE itself had a subsidiary nearby, making refrigerators for the regional market just as it did in Canada. This plant, too, wanted to buy a rival's compressors–Embraco's.

I returned to Louisville and gave my interim report. The numbers were daunting. It cost MABG over \$48 to make each compressor. It cost Necchi and Mitsubishi between \$32 and \$38. Sanyo, Hitachi, and Toshiba were designing plants that would make com-

Ira C. Magaziner is the founder and president of Telesis, an international consulting firm that is now a division of Cresap, a Towers Perrin company. He has published a number of books and articles, including Minding America's Business: The Decline and Rise of the American Economy *(Random House, 1983) and* Japanese Industrial Policy *(University of California-Berkeley Press, 1981). Mark Patinkin is a nationally syndicated columnist for the* Providence Journal-Bulletin. *"Cold Competition" is adapted from their book,* The Silent War: Inside the Global Business Battles Shaping America's Future *(Random House, 1989).*

pressors for under $30. Embraco and Matsushita's Singapore plant were aiming at $24–almost half GE's cost. One reason was labor. GE was paying over $17 an hour, including benefits–compared with Matsushita's $1.70 in Singapore and Embraco's $1.40 in Brazil. Even more astounding was the difference in productivity. It took GE 65 minutes of labor to make a compressor. It took 48 minutes in Singapore, 35 minutes in Brazil, and under 25 minutes in Japan and Italy. A company that's paying higher wages for lower efficiency doesn't have much of a chance.

The competition's export plans were equally daunting. Embraco was already shipping 10,000 compressors a month to the United States and aiming for ten times that amount within four years. Meanwhile, Necchi had just boosted exports to a million a year, and Matsushita would soon be in the multimillions. Overnight, foreign companies had gone from a small percent of the U.S market to a full 20%. And the real invasion had yet to begin.

> Overnight, foreign companies had claimed 20% of the U.S. market. And the real invasion had yet to begin.

MABG's biggest product was in jeopardy. If management didn't act soon, it could be disastrous for the whole group. The options? One possibility was to source. Another was to build a factory overseas in a low-wage country, perhaps in a joint venture. The third possibility was to invest in a new, more efficient factory here at home. It was clear which way Louisville was leaning.

"If they're that far ahead of us," one executive said, "how can we possibly catch up?" "We ought to just go for the source," someone else added. Even John Truscott, the group's head of engineering, wavered, shocked by the difference in labor costs. Most felt it was a shame that GE was being outcompeted and a bigger shame to think about closing a factory; but a company's first mission is to survive.

Although this was only an initial overview, for many it was enough. The sourcing bandwagon began to roll. Don Awbrey, a Louisville general manager who had been put in charge of the compressor project, decided to speed up plans for a sourcing option. It was a good idea. Even if GE built a new plant, it would take years to get it up and running. Meanwhile, they'd need a bridge.

Compelling as sourcing was, however, there were still good arguments against it. Once you shut your plants, you're in danger of being hostage to your suppliers, many of whom, in this case, were also potential competitors. To source a product when you're a second-level producer to begin with–as GE did with microwave ovens–is one thing. To source the heart of your biggest product when you're the market leader is a much larger risk. The tide in Louisville was moving toward sourcing, but was it the right direction?

The ideal alternative would be to build a new U.S. plant that could make compressors cheap enough to undercut those built by the dollar-an-hour people abroad. Could a high-wage country do that? Theoretically, yes–through automation. But not with the same product design as the competition–in this case, reciprocating compressors. Brazil's Embraco had automated the recips about as far as they could go–and with low-wage workers. GE's only hope was a new design–the rotary. Because it had fewer parts, there seemed a good chance of making it quicker, boosting productivity. That was what Toshiba was planning to do with its rotary. Could MABG do even better on its own? I doubted it. It's hard to become the leader in a new technology that a competitor has already started to run with. True, GE had invented the rotary for air conditioners, but refrigerators were different; they worked a compressor much harder. But there was no need to do it alone. If GE could get Toshiba to help with either a joint venture or a technology license, it would have a better chance with less of a gamble. At least it was an alternative. But proposing it to GE headquarters in Connecticut would take more than rhetoric. It would take a detailed plan.

To explore the rotary option, I made an appointment with Tom Blunt. He wasn't pleased to see me. He'd heard about my interim presentation–and the way the sourcing bandwagon was rolling. He was sure the decision had already been made. But I explained about the rotary and asked if he could put together a plan for a factory. He nodded. That's what he did for a living–it wouldn't be a problem. I told him he needn't design it from the ground up. Instead, GE could work with Toshiba or another Japanese producer. Blunt's response matched his name. He didn't want to do it that way. If we gave his guys a chance, he said, they could outperform the Japanese. But he doubted management would let them try. A new factory wouldn't be cheap, and back then, it was almost unheard-of for a U.S. smokestack company to fight foreign competitors with a big plant investment. Still, he said he wanted to do it–even though he didn't think anything would come of it. I agreed to back him. We'd propose building a new plant without the Japanese.

John Truscott was also intrigued. Before coming to Louisville, he'd pushed technology forward at each step of his career, first on an aerospace team involved in breaking the sound barrier, then when he'd helped perfect medical CAT scanning. Now he saw that same promise in the challenge of automating smokestack America. It was time to show that the United States could still be a world leader in manufacturing.

Truscott spent some time studying the rotary with an engineer's eye. He found it could indeed be made simpler than the old-fashioned reciprocating compressors. He also found that even the Toshibas were far from perfect. There was room to take this technology beyond the competition. He assembled a team to create a new design. The challenge was to make the compressor as simple as possible, with low noise, high efficiency, and – hardest of all – durability. At the same time, it couldn't be loaded up with too much metal or it would cost too much.

After a few months of work, the engineers came up with a model they were convinced could be made more cheaply than the one the Japanese were producing in Singapore. There was only one problem: the design required the key parts to work together at a friction point of fifty-millionths of an inch – about one-hundredth the width of a human hair. No product on earth had ever been mass-produced at such an extreme tolerance. Most engineers thought technology hadn't advanced far enough even to try. Tom Blunt knew some machines operated at those tolerances – jet engines, for example. But their parts had to be tooled one at a time, over long hours. Was it possible to get such precision in a plant that made 3,000 pumps a day? Everyone they talked to doubted it could be done. The design engineers took the idea to Truscott anyway. "It looks possible," Truscott said. He told Blunt to gather the people necessary to plan a factory. Blunt knew he'd just bargained himself into an obsessive few months. Designing a new factory is an enormously complex job with 100 new headaches a day. But that's why he likes doing it, he says. "Because it's hard."

Blunt knew this would be a first – one of the world's most automated factories. To design it, he'd need 40 people. Many GE colleagues advised him to go outside. To pioneer new technology, they said, you have to find designers who are already on technology's edge. But Blunt decided to stay with his own people. He gathered many from an unlikely place – Building 4. "We didn't go out and get a bunch of Star Wars guys," he would say later. "Most of these people came from one of the most nonautomated places you've ever seen in your life."

Why did he risk that? Blunt is convinced that American industry doesn't have to recruit experts for breakthrough projects. Most seasoned engineers can do it, he says. All they need is the backing – and confidence. At MABG, he knew his people had neither. "Some of the engineers here were the brightest people I had ever seen," he recalls. "They had degrees coming out of their ears. But they'd never been allowed to do anything." For years they'd been free to innovate on gadgetry but not on basic manufacturing. That left most of the engineers in a deep malaise. What made it worse, said Blunt, was that they were treated like second-class citizens. "A lot of people thought that we couldn't walk and chew gum at the same time."

If the engineers were to come up with a world-class breakthrough, they had to believe they could do it. So Blunt started by working on morale. As the engineers began their work, he pep-talked them. The reason he'd plucked them from Building 4 was that he needed people who knew factories – and still believed in them. He was convinced that they could design a better plant than anyone in Japan or Korea. True, no one had ever built a factory that could mass-produce parts of this precision – or achieve interchangeability at fifty-millionths of an inch. But none of that mattered. Here in America, in Louisville, they'd be the first. One of Blunt's favorite approaches was to remind his team that few outsiders would understand why they built factories for a living. You get no credit for it, he'd say, even though it's about the most difficult thing there is to do. But that's why they'd picked it, because of the challenge. And then he'd deliver the clincher: "*Anyone* can source," he'd say. Gradually, his people began to feel more confident than they had in years.

> It was time to show that the sun wasn't setting on U.S. manufacturing.

The problem with most manufacturing, Blunt felt, was that the factories were designed around products. This time, he and Truscott decided to design the process and product together, adjusting each as they went. They began by moving the product engineers and the manufacturing engineers across the hall from each other. Day by day, more and more people crossed the linoleum. Gradually, they fine-tuned the pump down to the most automatable model – a stationary vane rotary. With fewer than 20 parts, it was also the simplest. The computer simulation said it would work, but only if the machining went far beyond anything the Japanese were doing with their plants. To help find a way to do that, Blunt brought in specialists from GE's jet-engine division. He brought in computer modeling engineers he'd met at Ford. He brought in the head of the Swiss Institute of Technology and consultants from the Structural Dynamics Research Corporation. But they were just for advice. He still relied mostly on his own people, the people from Building 4.

Blunt's main rule was never to allow someone to say it couldn't be done. "We figured it was the way to drive our people beyond state of the art," he says. "If you say it can't be done, you won't do it. But if you say, 'We don't care that it's never been done, we're going to be the first,' then you have a shot."

> Tom Blunt's other rule: "Every piece of equipment must be made in the United States."

Slowly, week by week, the plan came together. "There weren't great 'eureka' breakthroughs," recalls Blunt. "It doesn't work that way. The whole thing was block-and-tackle grunt work." As he expected, there were 100 headaches a day. Constant frustration. Late nights. Blunt hadn't enjoyed work this much since he'd come to Louisville.

The factory began to take shape on paper. Each time they finished roughing out a new piece of it, they taped it into a matrix unfolding along a corridor wall. The matrix soon took up a quarter block of space. To keep it going, they had to find empty offices and extend the paper in there. They spent a lot of time just sitting, drinking coffee, and looking at it. How to integrate the grinding and the gauging? The loading and the material handling? They moved around pages, deciding what to automate and what to do with workers.

And then it was done. "But that still didn't mean anything," says Blunt. "It was just a bunch of sheets of paper. Anybody can do that." Now came the second stage. Could they design machines that would make the parts the paper called for?

Blunt had one other rule. He wanted every piece of equipment in this plant to be U.S.-made. His stated reason was that it's too hard to deal with vendors 12,000 miles away. But there was another reason. He wanted to show that the United States could outdo the world using only its own resources.

One of Blunt's chief engineers was Dave Heimedinger. Under his direction, a team of engineers began to negotiate with suppliers of grinding and gauging machines. The vendors would look over the plan to mass-produce parts at jet-engine precision, and then they'd shake their heads.

"You can't do that," one vendor said.

"We think we can," said Heimedinger.

"Well," said the vendor, "it's our equipment, and we don't think it'll do that."

"We think we can find a way to make it do it."

"Well," said the vendor, "all right. We'll sell it to you. But it won't do that."

One manufacturer insisted on putting a clause into the sales agreement saying the purchasers had been warned they would not be able to get the tolerances they hoped for. He also added a no-return clause. Heimedinger bought the grinding machine anyway.

Blunt's theory of how to make machines do something they weren't built to do was simple enough. "We played around with them." Heimedinger and his team began experimenting with combinations that had never been tried. Often people would come by

GE asked workers to sacrifice 120 to 400 training hours without pay. Workers lined up.

and ask Blunt why he was bothering. MABG is in trouble, they told him. Why gamble on a new factory? Let's just source and get on with it. Blunt smiles at the memory. "I just put my head down and said, 'Well, we're working on the damn thing.' "

The first prototype machines began to deliver parts the manufacturers had said they couldn't produce. Blunt knew it wasn't final proof. Test machining is a good guide, but when you're building an unprecedented manufacturing process, the only real test is the plant itself. "A first-of-its-kind factory is its own prototype," Blunt says.

On one hand, he considered that a welcome risk, a sign of limitless potential. If they could have proved the plant would work, it would have meant they weren't breaking new ground. On the other hand, it made for sleepless nights. Until it was done, and the switch was thrown, there was no way of knowing whether they'd succeed. Blunt did, however, have two tests he used to gauge whether each of the new ideas would work. One he calls the eye test. "If you look into an engineer's eyes," he explains, "you can see whether he feels good about something, or whether he's afraid of it." The other is the "I'll try" test. "If you hear an engineer say, 'I'll try,' " he says, "you better look real close, because he's afraid it's impossible." When they'd finally finished their plan, no one was saying, "I'll try."

One thing made Blunt more nervous than the technical question–the financial question. Would this plant make cheaper compressors than anyone else in the world? It was up to me to make those first projections. Both Blunt and I knew that if they didn't add up, it wouldn't matter how brilliant the design was. Headquarters couldn't possibly go ahead with it.

There was no getting around the fact that the costs were enormous. The plant itself would cost $120 million. And it would take tens of millions more to redesign the refrigerator so the new compressor would fit it. That would make it one of the biggest single investments GE had ever made in a factory. It was a lot to gamble on the hope that GE could produce cheaper goods–with $17-an-hour labor–than rival factories paying less than $2.

> When you're building an unprecedented manufacturing process, the only real test is the factory itself.

The risk was beginning to loom as almost unacceptable. Then GE's engineers came up with a way to reduce it. GE had a Columbia, Tennessee plant that made air-conditioning compressors–rotary models. Instead of going from nothing to the prototype factory in a single leap, they could begin by adapting the machinery already there. It would let them move into rotaries faster and work out the bugs while the new factory was being completed. The numbers on that proposal looked better. Even at ten times the wages, the new factory would still be the lowest cost compressor plant on earth. At least on paper.

Blunt knew this was no guarantee that GE's Connecticut headquarters would back the proposal. Fairfield was wary of huge capital investments, especially since GE had recently lost money on a failed washing-machine plan. How could MABG convince management to invest an even bigger amount on an even riskier venture? Especially when it could source for almost nothing?

I finished my calculations and called Blunt. He asked what I thought. Until now, I'd not come down firmly on either side. Now I said I'd weighed the two options carefully and had decided to recommend that GE invest the $120 million. "You really are a crazy son of a bitch," he said.

I went to Don Awbrey, head of compressors. After a thorough discussion, he called his people together and told them he was prepared to go with the factory. But he wanted an airtight proposal. So did Jim Lehman, a finance man who had been with GE for 30 years. Like all good finance people, he treated requests as if the money would come from his own pocket. At first he was doubtful. He made us recompute the numbers with every possible risk factored in. But finally, everything seemed to add up.

That's when the completed plan went to Roger Schipke, the new head of MABG. In his prior post as head of dishwashers, Schipke had successfully managed one of the few MABG projects in a decade that went beyond bells and whistles—a factory and product redesign that cut costs, improved quality, and doubled market share. This didn't mean that Schipke favored big investments, though. Conservative by nature, he had come up through sales, knew the importance of profit, and, like most in Louisville, had gone beyond the early 1970s mind-set about manufacturing everything in America. But one of Schipke's priorities was to get rid of the Louisville malaise. When a division becomes lackluster, people tend to turn on each other—Schipke was working hard to change that, getting rival managers to work together and stressing cooperation with the unions. Slowly, MABG was becoming more cohesive.

And now Schipke had been handed this proposal. The compressor team had numbers proving the new factory would be more than ten times as productive as any other. Because the plant was so automated, labor costs weren't nearly as big a factor as Schipke had expected. He said it was time to take the proposal to Jack Welch, GE's chairman.

Schipke, Truscott, and Blunt flew to Connecticut to make the final pitch. Later, Blunt would recall the plane ride. The betting was that they wouldn't get approval. A few years before, management had been almost haughty about the Japanese; they could never touch GE's quality or technology. That had changed. The attitude now was that the Japanese had become manufacturing geniuses. Why try to beat them when you can borrow from them? Or buy?

It was Blunt's first time in the Fairfield boardroom. The room was nearly empty as he filed in with the Louisville team. Then a half dozen executives from headquarters, including Jack Welch, came in. The Louisville people made their case.

"Jack poked at us three or four times," remembers Blunt. Then the chairman asked his colleagues their opinions. A few said that they doubted it could be done. Welch looked at Blunt. "Why should I believe you people can build a factory to do this?" he asked. "You've never done anything like this before."

"No one ever asked us to," said Blunt. "And I believe we can do it."

Welch nodded and turned to Ed Hood, a vice chairman of GE and one of his most trusted technical advisers. Blunt watched as he drew out Hood's comfort level. Blunt was counting on three things. Welch, he thought, had faith in Schipke's new management team. He'd seen the numbers showing the plant could do it—if the technology worked. Finally, the chairman wanted to keep major appliances as a core business for GE and was concerned enough about Louisville's slide to know that only big investments could turn things around.

Welch turned back to Schipke, Truscott, and Blunt. "Okay," he said. "Go ahead."

Keith Moore, recently transferred from GE's lighting business in Cleveland, was in charge of the startup. That first meant retrofitting the old Columbia, Tennessee air-conditioner-compressor factory with the new processes developed by Blunt's engineers.

Moore's people soon found that it's easier to design a new process than to make it work. The suppliers' warnings proved true. At first, GE couldn't make the equipment do what Blunt's engineers wanted. It took endless hours of debugging and hundreds of changes on each machine. The required tolerances were so extreme that even the tiniest slippage could throw a whole process off. Ultimately, GE had to develop new gauging and sensing systems to get the machines to readjust instantaneously as they worked.

Machinery deliveries were 2 months late at the start and up to 14 months late at the end. Management found it hard to shepherd the process from Louisville, 200 miles away, so GE rented 22 apartments in Columbia to house engineers. The company even started a daily air shuttle between the two cities so that the lab results could be flown in from Louisville and test observations could be flown back from Columbia. By October 1985, GE had finally begun Phase One. The old factory started producing the new compressor—first 5 per day, then 10, then 100. By month five, it was up to volume production, with the quality holding just fine.

> A few years before, GE management had been almost haughty about the Japanese. That had changed.

But if GE were to succeed at Phase Two—making the new fully automated plant work—it would have to face another challenge just as important as improving its hardware: improving its people.

A high-wage country can't compete with better technology alone; its other weapon has to be a better trained work force. Could MABG create that in a place like Columbia, Tennessee, where the biggest annual celebration is Mule Day? GE knew it would have to try. Hiring high-salary technicians from around the country was too expensive. At $17 an hour–benefits included–the new plant could still beat the competition, but not at $25 or $30 an hour. So GE planned to staff the new factory with the assembly people already there at its Columbia air-conditioner complex. Most were unskilled. Few had more than a high school education.

GE decided to make yet another big investment: it would build one of the most sophisticated blue-collar training centers ever put in a U.S. factory. The cost would be more than $2 million, which would have been difficult had GE not gotten help from a welcome partner–the state of Tennessee gave the company a training grant. But MABG still couldn't afford to pay workers for the hundreds of extra hours it would take to train them. So GE asked workers to sacrifice 120 to 400 hours in classrooms, labs, and computer stations without pay and with no guarantee of promotion–that would depend on how they performed. All GE could offer were new skills.

Paul Varner, who'd been named to help run the training center, thought the whole idea was a bad mistake. He'd worked on the Columbia assembly line and knew that most of the people there were conservative souls, wary of anything new. They already had secure jobs; what would be the point of sacrificing up to a year of nights and weekends for no pay? His guess was that almost no one would volunteer. "It took me two weeks to realize I was totally wrong," Varner says today. "I ate crow."

Workers lined up for the training. Partly, it was because of the prestige GE gave it. Those who made it through got diplomas and graduation dinners. But there was another draw as well, the same one that had prompted Varner to apply for a training-center job himself. He saw that plants throughout America were closing and knew it was a matter of time before distant forces put him out of work too. Columbia's old equipment was antiquated. They couldn't hope to beat 1980s rivals with a 1960s factory.

So when GE announced its new plant, Varner wanted to be part of it. He didn't mind unpaid nights and weekends in the training center. For him, joining the future was incentive enough. And, as it turned out, it was enough for hundreds of others too. Clayton Russell was among the first.

Russell had been hired in 1974 for an unskilled assembly job. "That's all we had then," he says. His job was to put four screws into the rear case of an air conditioner–712 times a day. Gloria Anthony began

Standing before their computer terminals, the line people are symbols of the new American blue-collar worker.

the same year, also on the line. "A monotonous job," she says. "Over and over and over." Then construction began on the new plant. To be a part of it, they'd have to put in hundreds of hours of training, all on their own time. It didn't matter. They attended training sessions mornings, nights, weekends. "Whenever we had a chance," says Russell.

Dan Edlin, another line worker, put in 400 hours. Like the others, he was motivated by more than the chance of a bigger paycheck. "I wanted the opportunity to be in on something totally brand new," he says. "This is where business is going–automation." Was there resentment that automation would cost jobs? "Machines aren't taking people's jobs," he says. "Machines are making new jobs. Anyone who wants to get off his duff and train can have one."

In the center's first year, the workers of GE Columbia spent over 50,000 hours learning new skills. Paul Varner learned a lesson: give American workers an opportunity and they'll sacrifice for it. "From Welch on down, they were saying, 'You can do it,'" says Varner. "We wanted to prove their faith in us."

Soon Keith Moore was facing his next challenge: moving manufacturing from the converted air-conditioner plant into the new factory. He knew that making the just-completed factory perfect would mean thousands of adjustments. He also knew that his floor workers would be best able to spot many of those adjustments. So, right from the start, he held meetings with workers and engineers sitting shoulder to shoulder to discuss getting better quality and efficiency. Moore was just as likely to reorganize a part of the plant at an assembly worker's suggestion as at an engineer's. And he soon got an unexpected payback from spreading responsibility through the ranks. In the past, if even a small thing went wrong, line workers had to call a supervisor to solve it. Now they fixed it themselves. Part of it, says Moore, is the training: they know what to do. But they also feel like they own their part of the factory–making it run is on them, not their bosses.

Moore also got workers involved in writing training manuals for the equipment. He figured that if they had to teach the techniques, they'd learn them better themselves.

Finally, he forbade finger pointing. Moore had seen before how workers blamed each other when things went wrong. So he announced that any failure would be considered the fault of the whole team. That way, Moore hoped, if one worker was having trouble, everyone else would rally. They did.

Looking back, Moore realizes how important it was to keep up camaraderie. Frustrations could have easily gotten out of hand. Late deliveries backed up the timetable. Processes that worked in the lab often failed on the factory floor. There were late nights and occasional 7 A.M. Sunday morning meetings. But they made it; the plant opened on schedule, in March 1986. Both production and quality went smoothly, though inevitably there were bugs. Having the older plant to occasionally fall back on was a blessing. GE admits it had to pour in more investment than expected–that happens when you're pushing technology's edge. But the factory is working, exceeding quality and cost targets. Ask Moore to point to one thing that did it and he won't mention hardware; he says it's the freedom he gave his workers. "We provided people with the tools to run their own businesses on the factory floor."

The celebration was short-lived. In January 1988, 22 months after the first compressor rolled out of the new factory, a problem surfaced. Some of the larger compressors–those in GE's bigger refrigerators–began to fail. It was only a small percentage of the plant's total production, but for a consumer product like this, reliability is essential and so is customer satisfaction. Schipke immediately formed a team of design engineers to find what was going wrong. The team worked for weeks, often through the night. The fact that only a small portion of the compressors had actually failed made the job especially difficult. But based on those few, the engineers went into a massive testing program and found that others could fail too.

Before long, the team discovered what was wrong: a lubrication problem was causing one of the compressor's smaller parts to wear more quickly than expected. The problem mostly affected the compressors that had to work the hardest, but some others were failing too. Eventually Schipke would learn that GE wasn't alone. Japanese companies using rotaries were having similar problems. But that was no comfort.

> In the past, line workers called supervisors to solve even the smallest problem. Now they'd fix it themselves.

Now that the cause had been isolated, finding a fix became Louisville's obsession. Truscott and some corporate engineers led a new team that worked on the problem for months. Finally, the team came up with a better design and showed it to Schipke. He was confident it would work and told the team to go ahead. Meanwhile, he approved a plan to replace immediately any compressor that broke down–service people would be dispatched to customers' homes at GE's expense.

Still Schipke faced a serious dilemma. While projections showed that only a small percent of the compressors would fail, redesigning the lubrication device would take months. He didn't want to risk GE's reputation by shipping refrigerators that might develop problems. To be as safe as possible, he made a painful choice. In the spring of 1988, he decided that MABG would start to source reciprocating compressors from abroad while the engineering team put the fix into place. It would mean a layoff at the old plant in the Columbia complex. And it would mean a high cost burden, since GE had to pay top dollar for the sourced recips and take longer contracts than it needed. Schipke knew it would probably cost GE far less to stay with its rotaries and just replace those that eventually broke down. But every failure, he feared, would hurt the company's reputation. That was more important than today's money.

Despite problems, the payoff stands: GE's compressors are 20% cheaper than those of its dollar-an-hour competitors.

He took some comfort in knowing he could let the new factory keep making rotaries for many GE refrigerators. He was also relieved that the group would be back to full production within a few years. But it was–and is still–a painful time. The compressor problem has drawn hard criticism from both the competition and the press. Some at GE are embarrassed because they had to source. Others are irritated at how much money the problem has cost: MABG's 1988 profits were way down because of the compressor problem, which has eaten up years of savings from the rotary program. Still others are angry because it was potentially avoidable. Early lab tests showed the compressors would last 20 years, but obviously, the truest test is performance in the field. Looking back, many MABG executives see a lesson. When using a brand-new technology, it may be wiser to introduce it gradually, working out the bugs over a few years, than to convert your entire production to it immediately.

The irony is that over 90% of GE's compressor investment–and risk–was tied up in the factory, by far the most complex technological challenge. And the factory works perfectly. A relatively simple part of the product design itself caused the crisis.

As GE knows, occasional product recalls are part of the price of gambling on new technologies. It happened with fuel-injected car engines, electric shavers, and microwave ovens when they were first introduced, and now it has happened with rotary refrigerator compressors. The cost to GE is high. But the payoff still stands: here in America, GE continues to make compressors that are 20% cheaper than any made by dollar-an-hour competition.

Clayton Russell, who used to put four screws into the rear case of an air conditioner 712 times each day, now runs a $700,000 synchronous machine with 12 different stations. Gloria Anthony is now a skilled controlperson who can operate machines by computer, adjusting them whenever the terminal tells her there's a slight problem. "I never thought I'd go this high," she says. "When I began, I was just sweeping the floor." Both are proud to be part of a plant that makes twice as many compressors as the old one, with less than a quarter the people. Productivity, they say, is the only way America can compete. "We got production," says Russell. "That's a feeling of pride."

Edward Fite, director of the training center, guides visitors around the automated Columbia plant with the pride of someone showing off a new home. Overhead, compressor pieces roll down long, winding chutes into machines that stamp, cut, and refine; computers direct the pieces from one machine to the next, warning that they're on the way. The machines work and work, never stopping. Grinders, welders, testers, and robots do their drilling, milling, tapping, and gauging. There may be no other mass-production factory in the world that makes goods this precisely. Most of the line people stand before computer terminals: they're symbols of the new American blue-collar worker, equipped with tools to outcompete the world.

At 36, Fite himself has come from stuffing wires to teaching workers how to run a high-technology plant. He sees people like himself as the final mission of a factory like this–it gives common people a standard of living beyond what they ever hoped for.

Tom Blunt will always be a figure from an earlier America. "I like the gritty stuff," he says. His father was a toolmaker, and he proudly traces his ancestry back through 11 generations of factory mechanics. But he knows the United States can lead the world in factories only by building a new kind. At his desk in Louisville, he leans over a computer terminal and punches a few keys. A moving diagram appears on the screen. He explains that he is monitoring the Columbia factory. Sitting in Louisville, 200 miles away, he can peer by computer into the guts of any machine he wants–judging how it's working, how many parts have been made correctly, how many had to be set aside for rework.

"Let me show you something interesting," Blunt says. He punches a few more keys, then leans back, clasping his hands behind his head. "Since 7 A.M.," he says, "we've made 3,413 pumps." He pushes an update button. "Sorry," he says, "3,415." He pushes the update button again. "3,417. So far," he says, "we've had only five defective parts today." He pushes another button and nods; the computer is showing him a diagram of the day's tolerances for one of the machined parts. "The problem was with one of the grinders that went down," he says. "It's fixed now." Is there any other plant in the world that has this automated an information system?

"There isn't," says Blunt.

Building 4 still stands a few hundred yards from Tom Blunt's office. Only now, it's half empty. Even if Columbia hadn't been built, it would have been dead by now. The choice was simple: it would either be replaced by a foreign plant or by an American plant; a foreign payroll or an American one. Despite the setbacks, GE is proud it chose to go with this side of the ocean.

Reprint 89211

GETTING THINGS DONE

Four basic problem-solving loops can help the most crisis-driven organization turn its operation around.

Fix the Process, Not the Problem

by Harold Sirkin and George Stalk, Jr.

In 1983, the CEO of a paper company faced a difficult decision. His board of directors had just met to consider alternatives to filing Chapter 11 for a subsidiary, a paper mill acquired two years earlier that was losing more than $1 million a month. The acquisition had been made to grow the company, but it now confronted management with the prospect of a major write-down. The price of the company's shares had already fallen 40%.

One year later, the mill was just about breaking even. Today it is very profitable. Earnings per share for the entire company tripled from 1983 to 1989, adjusted for stock splits, while price per share rose almost tenfold. Many of the mill managers can look forward to a comfortable retirement based on the value of their stake in the transformed company.

What happened? The short answer is that everyone at the paper mill became a problem solver. Together, managers and mill workers learned to take the initiative not just for identifying problems but also for developing better processes for fixing problems and improving products. Their approach did not depend on key senior executives taking charge and telling people what to do. Instead, the entire organization learned how to learn. The key to their success: a multiyear learning process in which employees developed four progressively more sophisticated problem-solving loops: fix-as-fail, prevention, finding root causes, and anticipation. (The exhibit "How the Mill Solved Its Problems" illustrates these four loops.)

Setting the stage

For the company's CEO and his senior managers, starting the learning process was neither easy nor obvious. It was simply the only alternative that could give the mill a chance.

In the two months following the board meeting, the management team made several hard but necessary decisions. It shut down a small, inefficient, costly pulp mill and three paper-making machines. It laid off roughly 25% of the work force, delivering a serious blow to the small town in which the mill was the largest employer. It fired the mill's general manager. But these actions, management knew, could only slow the mill's losses. At best, it would buy one year's breathing time to turn the operation around. Success would depend on shifting the mix of products sold by the mill. And at first that seemed an insuperable problem.

Of the 13 product lines the mill produced, only 4 were profitable or had the potential to become profitable because of the mill's special manufacturing expertise and equipment. The rest of its products made little or no money and were nothing more than filler to keep the machinery operating. Shutting down three machines had eliminated some of these undesirable products, but those machines had accounted for only 20% of the mill's production capacity. That left significant volume with limited profitability, but it could not be eliminated immediately if the mill was to sustain even a scaled-back overhead level.

Clearly, the mill had to expand its volume in the four attractive product lines fast. Cutting prices to attract business was out because it would destroy any potential for profitability. That left improving quality and service–areas in which the mill's reputation was conspicuously poor. (The mill had always had problems with quality and meeting delivery dates. Adding to the difficulty was the fact that the mill had sustained a long, nasty strike that had ended only a few months earlier. During the strike, the CEO discovered on his

Harold Sirkin and George Stalk, Jr. are vice presidents of the Boston Consulting Group. Mr. Sirkin often writes on world-class competition. Mr. Stalk's first article for HBR, "Time–The Next Source of Competitive Advantage" (July-August 1988), won the 1988 McKinsey Award and was the basis for Competing Against Time *(Free Press, 1990), written with Thomas Hout.*

way home one night that his brake lines had been slashed. Management's relations with the employees were adversarial at best.)

Management's first approach to the mill's problems was top-down. The senior managers met, analyzed the problems—as seen through their own eyes—and decided on ten key actions the mill would have to take to survive. Many of these actions required large amounts of capital and additional people, however—resources the company for the most part could not afford and that would not be available from the parent in the quantities needed. Moreover, the plans focused on solving specific problems, often in a slow and very costly way: for example, rebuilding two of the mill's paper-making machines to correct paper-thickness inconsistencies across the width and length of a sheet. The rebuilding would take two years and cost over $23 million; whether it would actually solve the problem was not clear.

The moment of truth came when the CEO's controller took the plans key managers had submitted and laid out the overall timetable for implementing them. The result startled the management team. Reaching break-even would take at least five years. With losses mounting daily, that was time the company did not have. Management had to find a way to improve results more quickly. After further discussion, the executives identified four approaches: add resources; fix fewer but higher leverage problems; fix problems faster; and learn from each experience to make subsequent efforts more effective. But the only way to make enough resources available was to enlist the entire organization's help to improve the mill's quality and service levels.

The idea of turning to your people and asking them to solve business problems sounds less bold and risky now than it did in 1983. But even so, knowing something has worked for others and moving onto unfamiliar ground yourself can be very different things. In this case, the CEO and the mill's new general manager began by taking their description of the mill's problems to lower and middle management. They discussed the mill's difficulties frankly and outlined what the product focus would have to be if the mill were to survive.

> Bonuses—for the manager and janitor alike—depended on overall profitability.

Making the stakes clear, the top managers pledged their commitment to the mill and made it tangible in important ways. They promised that the parent company would make capital investments when needed, despite the possibility of failure—a critical point, since the mill had not seen capital in many years. They promised to be available whenever called and passed out their home telephone numbers. They established a bonus plan for all the employees—from the mill manager to the janitors—that was pegged to improvements in individual product-line margins or volume but contingent on the mill's overall profitability. (For bonus purposes, the mill's financial performance would be assessed after six months and at six-month intervals thereafter.)

On the other hand, the downside was just as clear. Employees who weren't committed to saving the mill were invited to leave. (None did.) And the CEO left no question that the mill would be shut down if

How the Mill Solved Its Problems

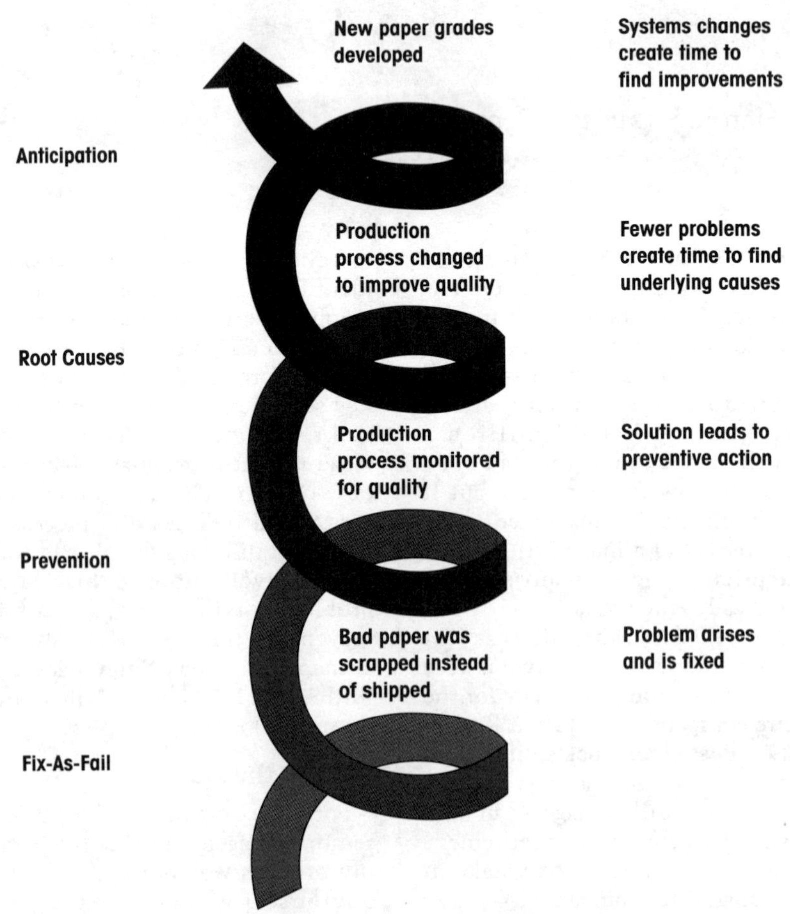

The mill worked through the four loops, progressing from problem solving on a fix-as-fail basis to anticipating problems before they arose.

this effort did not succeed. Then, knowing that most of the employees trusted lower level management, he asked those managers to get the message to everyone at the mill and to solicit their help.

The CEO put the company jet at the teams' disposal.

At the same time, in an effort to narrow the field of problems needing attention, the CEO implemented a new organization structure designed to make it easier for everyone to focus on problems that mattered to customers, instead of on individual functional goals. He created five product teams, one for each of the four attractive lines and one to manage the remaining products during the transition period. These teams were composed of seven middle managers, one from each of the key functional areas, and were headed up by the product-line salespeople (to reinforce focusing on customers). These managers were charged with enlisting the help of all the employees in their areas and urged to give everyone a chance to contribute.

To keep the teams from making the same mistake management had almost made – focusing on the immediately obvious problems and trying to solve them all at once – and to find further opportunities to narrow the list to problems with the greatest leverage, the CEO suggested that each product team take the next few weeks to talk with key customers to learn their needs and develop plans for responding. Beyond this, he let the team managers set their own agendas. He simply reminded them from time to time that they always had direct access to him and the general manager. And he reiterated his promise that resources would be found. To signal his commitment, he put the corporate jet at their disposal and used commercial flights for his own visits to other operations.

Learning to learn

In the weeks that followed, each team expanded from its seven-manager core to draw in line operators, clerks, and other employees whose work related to its product line. To give the teams time to meet, the CEO created time – by authorizing everyone to ignore the crisis and stop fighting fires. Predictably, the first month was a disaster, with problems piling up and seemingly nothing to show in return. But before long, management's faith in the learning process began to pay off.

To illustrate what this process entailed, we will concentrate on the activities of one team, the food-service product team. For the turnaround to succeed, its products had to become the mill's largest line, doubling the existing volume. At the start, this looked like an impossible goal. Food-service product lines got more customer complaints than any other product, and they had gradually been losing market share.

Within ten days of its formation, the team had met with its three largest customers, all of whom rearranged their calendars to accommodate the team's schedule because they so badly wanted a reliable source of paper. These customers told the team in detail about the consequences of the mill's poor service and quality problems. Late deliveries could shut down their plants, so they compensated by holding large amounts of inventory and purchasing paper from other suppliers. Quality problems, such as holes in the paper, ruined their own end products, so they incurred additional costs inspecting incoming paper and also bought from the mill's competitors. These discussions highlighted how shipping poor paper cost the mill business. The team returned to the plant determined to solve these and other quality problems.

Team members had now seen and heard their customers' needs firsthand. They had talked with equipment operators, salespeople, and receiving employees as well as with the buyers. And they learned things they never had before – for example, the effect on a customer's operation of the mill's paper-winding procedures. Wound one way, the paper moved smoothly through the customer's equipment. Wound the other, nothing worked right. It was as easy for the mill operators to wind the paper one way as the other. But before, no one had ever known it made a difference.

Back at the mill, team members talked with everyone in their individual areas and reported what they had learned in the customers' plants. More importantly, they enlisted everyone's help in resolving the problems and brought along people from all levels, including machine operators and shipping clerks, on subsequent customer visits.

Loop one: fix-as-fail

Operators at the mill knew all about the line's quality problems. But poor quality paper had always been shipped to make tonnage goals set by the CEO and agreed to by the mill manager. The team decided on its own to do its best to stop shipping defective rolls, whatever the cost of rejecting paper. A team member checked every roll and, if there were problems, set it aside. To show customers that quality mattered, the team also asked the mill's general manager to be the final inspector and sign every roll. At first, half the rolls were rejected and later salvaged by removing the defective portions. This salvage process was costly and time consuming, but customers were not sent low-quality rolls. The team had developed its first important problem-solving loop.

Getting stuck in fix-as-fail problem solving guarantees limited productivity gains.

Put simply, fix-as-fail means that when something goes wrong, the product or service is fixed before it is sent to the customer. It is the most basic kind of problem-solving loop, one that every company engages in to some degree or it could not remain in business. But many organizations get stuck in a fix-as-fail problem-solving mode, guaranteeing static performance and limited gains in productivity. The frequent need for last-minute fixes, the unpredictability with which problems crop up, the recurrence of the same old problems – all lead to stress, extra hours, and finger-pointing throughout the organization. Production processes come to seem uncontrollable. Operations often fall further and further behind with no apparent hope of catching up, since problems seem to accumulate faster than they can be solved. Improvement looks – and is – impossible because an ever-growing number of resources must be thrown into dealing with continual crises.

Loop two: prevention

To avoid or escape a fix-as-fail cycle, an organization must be able to engage in a second kind of problem-solving loop: developing processes to keep problems from occurring. For food-service product team members, this loop began when they looked at the buildup of scrap from rejected rolls and realized they had to find a way to stop solving quality problems through inspection. Realizing that they had to walk before they could run, they agreed as a first step to assign an inspector to watch as each roll was produced. When a defect appeared, the inspector would tell the machine operator, who could adjust the equipment immediately and prevent further problems. Adding the extra person was expensive, but productivity (as measured by good output per person) improved measurably. Scrap rates dropped dramatically, the quality of the product remained high, and the reject rate fell by some 75%.

As this example indicates, prevention does not depend on tracing problems to their root causes but on developing learning loops to feed information rapidly from the point where problems can be discovered to the people who can keep them from recurring. At one level, therefore, second-order learning is simply a matter of resource allocation. Investing one person in increasing the effectiveness of this problem-solving loop, as the paper company did, can free up 10 or 15 downstream people, who had been correcting problems on a fix-as-fail basis, and give them the time – and motivation – to embark on the third stage of learning, finding root causes.

At another level, however, second-order learning is also an organizational issue. Unless they are tackled head-on, old habits can enter in to keep people from moving beyond fix-as-fail problem solving. The corporate culture may block change. Employees may not have the tools to find and institute an effective problem-prevention process. Everyone may have accepted problems as a normal and inevitable part of doing business – especially if they are routinely rewarded for solving rather than forestalling them.

> Six months out, customers were beginning to notice a marked difference in quality.

Loop three: find the root cause

Six months out, the team had the production process under control, and customers were beginning to notice a marked difference in quality. One of the mill's best customers called and said it would increase its orders if the quality level held. The team was pleased but not satisfied. Assigning someone to watch every roll was inefficient, and scrap rates remained much higher than desired. It was time to look for root causes, a process in which operators, not managers, would take the lead because they understood better than anyone how the machines worked.

Holes in the paper continued to be one of the team's biggest problems. By experimenting with the many variables that control paper quality, the operators and team members learned that they could produce paper with fewer holes if they changed the wire mesh in the paper-forming machine more frequently. They tested their finding, confirming that more frequent changes produced higher quality paper. Then they asked the mill manager if they could integrate this new procedure into the process permanently. He granted permission on the spot.

The operators also found that they could raise quality and lower the total cost per ton of good paper by using pulp furnish that was slightly more expensive. (Furnish is the mix of pulps that constitutes the raw material for a paper-making machine.) The team made this change too, without asking for approval from the mill manager until it could demonstrate the positive effect. Scrap rates fell further, and quality steadily improved. Before long, the team was able to do away with the in-process inspector because the machines were producing quality paper almost all of the time. The root causes of the quality problem had been found and eliminated.

By the end of the year, satisfied customers were giving the mill increasing shares of their business, the volume of the top four products had doubled, and profit margins had improved. The change in morale was equally dramatic. People knew they were winning the fight to save the mill. Friction between management and labor all but disappeared. Everyone knew more about the mill's operation and customers' needs. Customers who had previously been discouraged from visiting the mill were invited to tour the cleaner facility. The organization now felt strong enough to move from defense to offense.

Loop four: anticipation

One of the keys to the food-service product team's success was its new understanding of its customers' needs and how their businesses worked. Such understanding would have been unimaginable while the mill's people were spending most of their time fixing problems. Perhaps the most dramatic example of this understanding came when a machine operator's suggestion led to the discovery of an unsuspected competitive advantage.

As part of the team's effort to learn about one customer's business, the operator had gone to one of its plants and spent a long time watching the machines and talking with their line operators. He returned to the mill convinced that the customer could run slightly thinner paper just as efficiently as the paper it was then using, and he proposed that the mill offer it as an alternative. The team liked the idea, but the mill manager was not convinced, since paper is sold by the ton and thinner paper weighs less. Rather than reject the idea outright, however, the manager and the team reviewed the proposal with other teams, key functional managers, and key customers.

One floor worker's idea unearthed a competitive advantage.

The breakthrough came when the mill manager realized that their key competitor did not have machines that could produce lower basis-weight paper. If the mill reduced its paper's weight, it could provide a low-cost alternative that competitors could not match. The mill could therefore raise its price per ton and make more money, while the customer's cost would go down since it would need less paper (by weight) to make its product. Everyone was ahead except the competitor.

After this, operators were invited to work with the team to develop new grades of paper. With experience, the team found new combinations of furnish and additives that let them make lower cost varieties that they had not been able to manufacture before. The operators were also able to alert the team to potential manufacturing problems before the new papers were put into production.

The food-service product team's volume did not just double, it quadrupled, while the number of employees stayed the same. The mill went from being the fifth of five suppliers to being the industry's number-one supplier for its products in two-and-one-half years. By 1986, the mill had to expand the capacity of its machines to meet demand. The organization had moved from being a mill that faced shutdown into the lead mill of the group.

Despite their success, mill employees are healthily dissatisfied. They want to know more about their company, their customers, their suppliers, and their competitors. Team members and operators have purchased samples of competitors' products to study and analyze. They have also asked for, and are receiving, training in management techniques as well as in new skills. Formally, the organization chart looks the same. But informally the organization has been revamped. Operators move from one machine to another as necessary or where there is a problem. People participate and make suggestions without worrying about layoffs. Bonuses (averaging 10% to 12% of compensation) have allayed workers' fears that management might get more than its fair share.

Patience makes progress

One mark of a world-class organization is that its managers seem to have easy jobs. Operations flow smoothly, and people put more time and energy into making improvements than reacting to problems. Such an organization will typically spend 80% of its time on root-cause and anticipatory problem-solving loops. In contrast, a lesser performer is likely to spend 90% of its effort in the fix-as-fail loop.

Executives from lesser performing companies often return from visits to superior operations charged up to achieve similar results. In their enthusiasm, however, many overlook the fact that the effectiveness they admire comes from steady progress through increasingly sophisticated problem-solving loops. And that progress, as we have seen through the paper mill's experience, takes time. Although the mill began to see the effects of its learning in three to six months (a typical time frame, in our experience, for the first signs of progress to become apparent), it took two years to reach the fourth loop. Depending on circumstances, a company may need as many as five years to work through the entire cycle.

Given this, it is easy to understand why many managers itch to short-circuit the process and jump to root-cause problem solving at once. But most companies that take this approach achieve unsatisfactory results, both because the organization does not truly understand its problems or the processes needed to resolve them and because it lacks the necessary resources. Instead of better performance, the net effect is overworked employees who neither solve problems well nor do a good job of maintaining the day-to-day business.

A better, more systematic approach to organizational problem solving should begin with a few simple questions:

☐ What are the critical problem-solving loops in our company? For a manufacturer, for example, these might include engineering change notices and warranty-claim resolution, while a service company might focus on satisfying nonroutine requests and its response to customer complaints.

☐ How well does our operation work from the customer's perspective? Do we respond to problems quickly and well or are we slow and unreliable?

☐ Do we learn from problems or are we continually fixing the same problems over and over again?

☐ Where do we spend most of our effort: on identifying and responding to specific problems or on resolving underlying causes and finding new ways to improve?

Progressing through this four-loop cycle is both a natural and an unnatural experience. Moving from fix-as-fail to prevention demands a great deal of effort, but there is little subtlety about what needs to be done. The problems that need attention are clearly identifiable. The demanding task is getting people to organize around them, if only because something else will always seem more important to one or more of the critical players. By the time the organization is ready to move from root cause to anticipation, however, this balance typically shifts. By then, problem-solving processes are familiar, so everyone is likely to be much more at ease. But the problem-solving loops that will yield meaningful improvements to the business (as the operator's suggestion did at the paper mill) are much more subtle and often depend on finding creative ways to expose employees at all levels to the right people in customers' organizations.

As an organization establishes more sophisticated problem-solving loops and begins to learn, its fundamental culture changes. But unless management supports this change by rewarding new performance metrics as well as those it has traditionally used, no one is likely to stay the course. At the paper mill, for example, the CEO took a number of unusual steps to let people know it was safe to expand beyond the old ways of working. He rewarded early change makers by recognizing their accomplishments publicly. He promoted people who were trying hard to manage the right processes and to make the right decisions, even if their efforts were not entirely successful. Perhaps most dramatic, though, was the way he handled the mill's first six-month bonus review. By the numbers, the mill just missed making its profit hurdle. Performance had pulled up a lot, but not quite enough to meet the plan. Nevertheless, the CEO decided (and the board agreed) to pay maximum bonuses anyway to support and reinforce what was happening. They knew that customers' perceptions change slowly and blamed the shortfall on history, not on the efforts under way.

In the pressure to get things done, many managers fear being patient. They focus on short-term fixes to existing problems rather than on instituting processes to solve and eventually prevent problems and to identify unsuspected opportunities. But as in the fable of the tortoise and the hare, the companies that seem to move most slowly and laboriously at the start often lead their industries by the end of the day.

Reprint 90411

Conventional wisdom says: Don't use flexible machines on high-volume lines. Cummins's decision to bend this rule has paid off handsomely.

Cummins Engine Flexes Its Factory

by Ravi Venkatesan

The rules are changing. Are we? In 1986, Cummins Engine Company, a manufacturer of diesel engines based in Columbus, Indiana, was in a predicament familiar to many Rust Belt companies. Until the mid-1970s, Cummins had enjoyed a relatively stable product line and steadily held more than 50% of the market. We made what our customers believed was an indestructible engine, the NT series, and our factories were organized to accommodate the consistent demand for the engine. We ran enormous volumes off dedicated machining lines with splendid efficiency, as many as 500 engines a day – batch manufacturing raised to a science. Nobody at the top of the company worried much about our carrying huge work-in-process inventories so long as we delivered good quality on time.

Then came federal emissions regulations and, of course, intense competition from companies like Komatsu and Caterpillar. Cummins has since had to launch seven generations of the NT engine and three completely new engine families – the B, C, and L series. Meanwhile, customers have come to expect increasing attentiveness and unprecedented quickness of response to their individual needs. As if this weren't enough, Cummins has had to cut prices on new products from 20% to 40% just to retain market share. Inevitably, senior management mandated that the factory reduce costs and lead time – which is another way of saying reduce WIP – at the very time the factory was trying to contend with proliferating parts.

In 1986, our catalog offered well over 100,000 parts. And though our designers were becoming more sensitive to our manufacturing difficulties, the number was still rising. For one of our newest engine families, Cummins offers 86 different flywheels, 49 flywheel housing options, and 17 types of starter motors with 12 possible mountings – all of which could be assem-

> For one engine family, Cummins offers 86 different flywheels, 49 flywheel housings, 17 starter motors, 12 possible mountings. No wonder the catalog has more than 100,000 parts.

Ravi Venkatesan is a group manager at Cummins Engine's fuel systems plant in Columbus, Indiana.

bled in approximately 1,200 combinations to satisfy customer applications. So while we were struggling to make parts for all the new models – not to mention for that guy in Kansas who just loves his 1938 workhorse and stubbornly refuses a free replacement – we were faced with having to customize products, modify parts that did not prove out in the field, and deliver in shorter lead times. Markets were getting more fragmented, and demand was getting more cyclical. All of these pressures warranted lower volume production runs – not the best thing for unit costs. Worst of all, low-volume parts were clogging lines previously dedicated to high-volume parts.

Frustrated by the constraints of the old batch manufacturing, Cummins commissioned a small task force to develop fundamental insights into the problem of reorganizing our factories for a proliferated product line. We went to work in early 1987, studying other companies' factories as well as our own, and what we learned has worked out well for Cummins. But it has also given unexpected depth to the term "flexible factory." We exposed something important about how advances in the machine tool industry – as dynamic today as those in information technology – are shaping the new competition in durable goods. The rules, indeed, are changing.

Beyond Cellular Manufacturing

Most factories begin as job shops: the scale of production grows, machines are added, and so are new products. Before long, a "process layout" results, by which like machines are grouped together – lathes with lathes, grinders with grinders. Batches of parts shuffle from the milling department to the drilling department. There are many well-documented disadvantages to this mode of operation, not the least of which is the evolution of extremely convoluted flow paths. It takes enormous cumulative time for half-finished parts to move from department to department and then wait in line. The result is excessive lead times, bloated inventories, and no clear-cut responsibility for WIP.

It is precisely to preempt such problems that so many manufacturers – including Cummins – have reorganized their factories into product departments, each responsible for a family of related products. They've grouped their machines into "cells," U-shaped clusters that move a part from lathe to grinder to milling machine with little interruption. Dedicated to a more or less narrow range of parts, these machining cells are tended by just one or two operators; their logic is no more remarkable than that of the transfer lines of hard automation. In cells, WIP gets handed off from machine to machine without "resting" and thus goes through conversion processes smoothly, often with a simple pull system such as kanban to manage flow. Striking improvements almost always result from the move to cells. Many a manufacturing consultant owes his livelihood to the real savings in lead time, WIP inventory, and direct labor that can be realized by adhering to the just-in-time canon.

> Cells choke when they are constantly changing from one part to another.

Between 1980 and 1985, most of Cummins's factories had already made the transition to cellular manufacturing. At the Columbus plant, we had reorganized into 15 cells, each machining a comparatively small group of related parts: a water pump line, a flywheel line, a manifold line, and so forth. We invested only $60,000 for new machines, $105,000 for additional fixtures and tools, and about $40,000 to move things around. In the first year, we realized a 25% reduction in floor space, a 30% reduction in scrap, and a more than 90% reduction in WIP inventory. We saved well over a million dollars in labor costs (mostly indirect).

Yet one does not need the imagination of Frederick Winslow Taylor to see that cellular manufacturing techniques by themselves are inadequate to the challenge of a greatly proliferated product line – just the problem that had come to plague Cummins. The key to cellular manufacturing – as JIT experts such as Richard J. Schonberger write – is changeover. If a cell is going to machine, say, 7 parts efficiently, cell operators have to be able to change tools and fixtures in consecutive lathes, drills, and grinders fast enough that downtime does not seriously erode the gains (in time, labor, and so forth) of having organized into cells in the first place. But what happens when 7 parts become 14 and runs of 10,000 become 500? Machining cells choke.

Worse, as I've already implied, parts tend to proliferate at the extremes – the beginning and the end – of their life cycles. New parts change as R&D perfects designs; mature parts become more customized. And Cummins was committed to bringing out new products and to servicing old products with equal determination. The proportion of low-volume parts shot up even more. We found ourselves subject to the "20-80 rule": about 20% of the parts – the high runners – accounted for better than 80% of the production output, while the remaining 20% of production caused 80% of the disruption. We were not the only company with such disproportionate production figures. In one large industrial equipment company we studied, about 80% of the product variety accounted for a meager 5% of total production but consumed 25% of the total labor.

Fast Changeover Is Not Enough

Distortions of this kind have forced piece-part factories to make too many changeovers. WIP in machining cells does not flow smoothly unless it flows consistently. If, after machining a pallet of parts, every cell operator had to stop for several minutes (in some cases, hours) to change tools and fixtures, would it not have been more rational simply to stick with the job-shop process layout? Granted, cycle time is bound to be quicker in dedicated machining cells than in batch manufacturing. But total factory throughput can slow down to unacceptable rates. It made no sense for Cummins just to keep adding parts to machining cells (and production departments) in order to exploit our equipment "to the full" or to retain hypothetical efficiencies of scale. What we had was a large number of different parts vying for the same production resources–and tying them up.

Predictably enough, some people in the company argued that we should simply work harder on reducing setup time. But close study of Cummins and other equipment manufacturers revealed that changeover success was restricted to higher volume manufacturing lines, that is, lines that typically produce fewer than ten parts. Moreover, whittling away at setup would mean diverting a great part of managerial attention to a disproportionately small part of total production output–that is, to the low runners.

We determined that aggressive setup reduction alone was not enough to solve the problem of low-volume parts congesting dedicated lines. We began to develop a more subtle way of sorting through high-volume and low-volume parts–the latter being most disruptive to fast-cycle lines. Inevitably, this also meant looking harder at flexible machine tools, in spite of their slower cycle times. And as we proceeded to focus on volume–that is, *to focus the factory not only by product but also by volume*–it dawned on us that we held one of the keys to making flexibility work. After all, high-volume and low-volume systems must coexist in virtually every factory. Yet managing these two types of systems requires very different attitudes, skills, equipment, and quality principles.

Take the management of the work force. Managing high-volume processes calls for Taylorist notions of supervision and control–time-and-motion studies wedded to close monitoring of workers' coordinated performance. In low-volume environments, management emphasis is on planning and support –not supervision per se–and on maintaining and programming complex equipment. Low-volume environments also call for skilled manufacturing personnel who are familiar with a wide variety of products and associated processes.

Order processing, logistics, and production scheduling also are different. High-volume lines can usually be fed by means of pull systems. Low-volume lines, in contrast, require longer range, often computer-generated projections. They also require greater planning with suppliers, who need to deliver small lots of raw materials (castings, forgings, fasteners) not just in time but on very short notice. In fact, more than 50% of missed shipments in low-volume products at another major manufacturing facility we studied resulted from difficulties in order processing, material ordering, or production planning. We observed that, predictably, high-volume parts took priority over low-volume parts whenever they intermingled.

And consider the different approaches to quality. In repetitive manufacturing, statistical process control is used to chart the progress of critical product/ process characteristics. In low-volume manufacturing, SPC becomes more difficult because of the short batch runs. The emphasis here has to be on the control of variables such as tooling and fixturing. There must also be stringent process documentation to ensure replicability of manufacturing procedures.

Finally, we learned early on that if successful management of the flexible factory presupposed focusing the factory by both product and volume, classifying parts merely in terms of high volume and low volume was not going to work. The distinction was too coarse to be practical. Is it more disruptive to a machining cell to manufacture 5,000 pieces of a part that requires one hour for setup or to manufacture 10,000 pieces of another part that needs four hours for changeover?

Classifying Parts as They Are

As we probed the functional characteristics of the various parts that Cummins makes, we were able to discern four categories of engine components (see the illustration of the Cummins NT Engine for examples of each):

Skeletal components such as the block, head, and connecting rod – the heart of the engine. These parts are characterized by low proliferation, high volumes, and relatively stable designs.

Engine subsystem components such as the water pump and lubricating oil pump. These parts guarantee and optimize engine performance and are characterized by high volumes and low proliferation within the current generation of mature engines. For newer engines, however, designs can be quite volatile; there are frequent design modifications to enhance engine performance.

Application components such as exhaust manifolds and flywheel housings – parts that reflect individual customer preferences. These parts are highly proliferated but enjoy moderately stable designs.

Performance components such as turbochargers and compression brakes. These components are characterized by low proliferation, moderate volumes, and reasonably stable designs. This group also includes parts such as fuel injectors, which may require either high or low volumes but still demand frontier manufacturing methods such as electro-discharge or laser machining.

After digesting these distinctions, we developed five general classes of components using the three parameters of volume, predictability of demand, and stability of design: (1) high-volume products with predictable demand and relatively stable design; (2) medium- to low-volume products with predictable demand but rapidly evolving design; (3) medium- and low-volume parts with predictable demand and stable design; (4) low-volume components with very unpredictable demand; and (5) products requiring nontraditional manufacturing processes.

The last class of components, requiring specialized or unique processes, seemed impossible to separate out on a volume basis. All such components would flow to the same machining area, where they would have to coexist, regardless of demand. The machinery of, say, laser machining is too expensive and the technical skills too specialized to expect any Cummins plant to support more than one such production area. Parts would simply have to wait in line; high runners would go to the head of the queue.

Things got more interesting with the first four classes of parts. Consider a hypothetical factory that has made the transition to cellular manufacturing and where demand patterns are indicated in the table "What Are Disruptive Parts?"

Given its high volume and stable design, block A will probably be made on a transfer line. Since the changeover from block A to block B is relatively quick, B could be manufactured on the same line. However, the situation with water pumps is clearly different. Although water pump A is a high-volume part, an investment in dedicated or special-purpose machinery will probably not pay off given the part's short design life. At Cummins, high-volume products with volatile designs are increasingly made in flexible machining cells, as are prototypes and new product introductions.

Water pump B, which shares many design features with A but with lower volumes, is also a natural can-

The Cummins NT Engine

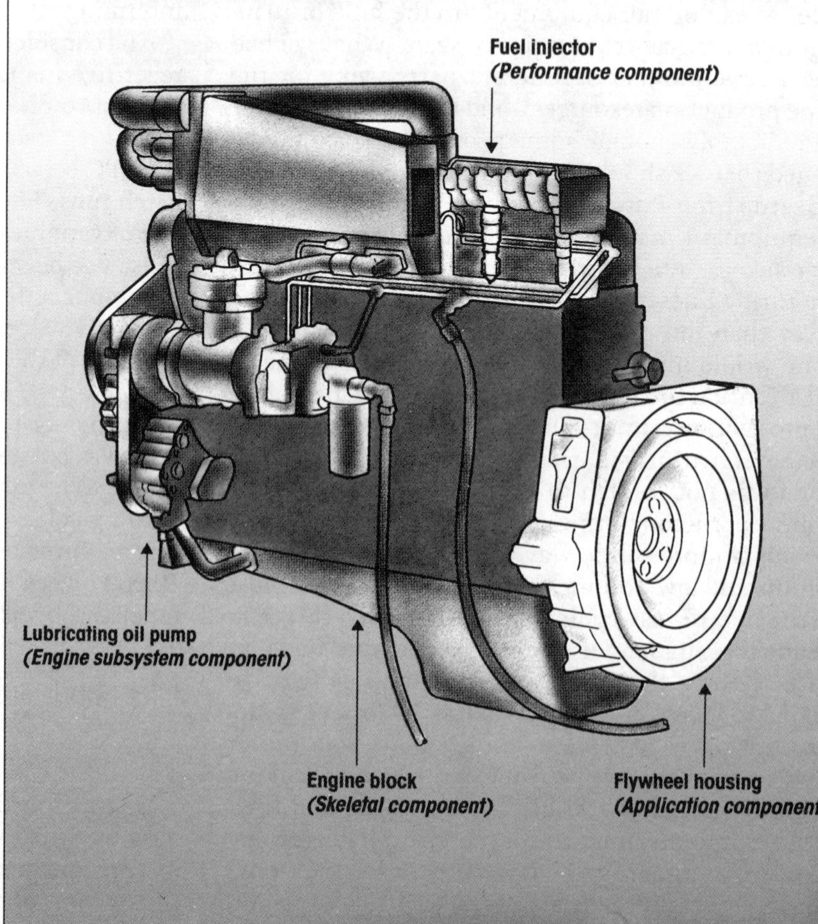

DRAWING BY LORI BAKER

What Are Disruptive Parts?

Disruptive parts are generally low-volume parts that interrupt the smooth flow of material by forcing frequent and lengthy setup changes on machines.

Product	*Annual production volume*	*Frequency of delivery*	*Design stability*	*Time to change over*
Block A	75,000	Daily	Stable	Changeover from A to B takes 30 minutes.
Block B	5,000	Monthly	Stable	
Water pump A	75,000	Daily	Two-year design life	A and B have common features, and changeover from A to B takes 8 hours. C is a very different pump. From A to C takes 2 days.
Water pump B	2,000	Unpredictable	Stable, obsolescent	
Water pump C	5,000	Daily	Stable	
Manifold A	75,000	Daily	Stable	Changeover from A to B takes 8 hours on a boring machine. From A to C takes 2 days.
Manifold B	5,000	Monthly	Stable	
Manifold C	20	Unpredictable	Stable	

didate for this same flexible cell. Setups that used to take eight hours now take less than one. The flexibility of the cell is wasted, however, on a part like water pump C, with its stable design, medium volume, and very predictable demand. The most economical way to produce C, we have observed, is to use existing equipment in a conventional cell, much as Richard J. Schonberger advocates in his article on frugal manufacturing.[1]

As for the manifold department, it presents several interesting challenges. Manifold A, the principal product, is an excellent candidate for a cell of dedicated machines, given its high volume, stable design, and predictable demand. Manifold B disrupts the smooth flow of material in this cell primarily because it takes operators eight hours to set up the boring machine on the same line. A good approach in cases like this is to create a "hybrid cell" with several conventional machines and a computer-numerically-controlled boring machine. Features common to all parts are produced on the conventional equipment. Features that differ from part to part are machined on the CNC machine.

There is no question that the flexible equipment will slow down the cell's cycle time. But setups are minimized. Overall, the throughput of the cell is quicker than it would be if the flexible machine were replaced by a special-purpose machine, which cranks out the product more quickly but needs much downtime to accommodate changeover. Hybrid cells can also handle such things as mounting hole patterns on flywheel housings, which do not change significantly over time. Most flywheel housings can continue to be produced on high-volume lines, while the variations might be sent off line to flexible machines. More and more production departments would need to acquire flexible machines to handle such variations; raw material and logistics systems would not be materially affected.

Finally, manifold C–the real killer. Its lengthy setup time and virtually unpredictable demand disrupts day-to-day operations. Products like C, we discovered, are best separated out into a very flexible, low-volume manufacturing area. Such products will be manufactured from start to finish in this production department, eliminating any ambiguity about who has final production responsibility.

To help us differentiate parts, especially low-volume ones, we developed a simple, rule-based "expert system," which appears in a simplified form in the table "Machining Option Adviser." Based on operators' answers to 14 questions, we can supply a quantitative recommendation about whether to machine a specific part in a conventional cell, in a flexible manufacturing system (FMS) cell, or on CNC machines in a low-volume area.

Take the example of low-volume part X. It has less than four faces, so it scores a 3 for a CNC machine, 3 for an FMS, and 1 for a cell of conventional machines. It is ten years old, so it scores 0 for CNC, 0 for FMS, and 2 for cell. After scoring for each criterion, the values for each machining option (some values are negative) are totaled up. The option with the highest score is the best way to manufacture part X. In the

1. See "Frugal Manufacturing," HBR September-October 1987, p. 95.

2. HBR November-December 1986, p.69.

Machining Option Adviser

Criterion		*CNC*	*FMS*	*Cell*
How many faces does the part have?	Less than 4	3	3	1
	More than or equal to 4	2	1	−2
How old is the part? (in years)	Less than 1	3	3	0
	1 to 3	2	2	1
	More than 3	0	0	2
How often has the design been changed? (number of design changes divided by part age)	Less than 1	0	0	2
	1 to 3	2	2	−1
	More than 3	3	3	−3
What is the cycle time? (in minutes)	Less than 3	1	−2	−2
	3 to 9	1	−1	0
	10 to 30	2	0	1
	More than 30	1	1	3
How many operations does the part require?	1 to 5	0	0	1
	6 to 15	2	2	−2
	More than 15	3	3	−3
Does the part need off-line operations such as heat treatment? *	Yes	−2	−3	−2
What is monthly demand multiplied by cycle time? (in days)	Less than 2	2	1	−2
	2 to 4	1	2	−2
	5 to 7	−1	0	0
	More than 7	−3	−2	3
Is the part disruptive to the current production process? (low volume divided by high setup time)	Yes	2	2	−1
Do you experience more than 2% scrap due to casting problems and variations?	Yes	3	1	−2
What is the tightest size tolerance of the part?	Less than 0.001″	1	1	2
	0.001″ to 0.003″	1	1	1
	0.004″ to 0.015″	1	1	2
What is the tightest geometric tolerance?	Less than 0.001″	2	2	1
	0.001″ to 0.003″	1	1	1
	0.004″ to 0.015″	1	1	3
Is the demand for the part unpredictable, with occasional surges?	Yes	2	2	−2
Is a significant reduction in cycle time possible through the use of multispindle machines?	Yes	−2	−2	3
If you had a real JIT operation, how many lots of this part would you run each month?	Less than 1	3	−2	1
	1 to 4	2	2	−1
	More than 4	0	3	−2

Values indicate the appropriateness of each manufacturing option with respect to various part characteristics, such as age and cycle time: 3 is the highest value, −3 the lowest. The manufacturing option with the highest total score – all 14 values added together – would be the best one to make the part.

* On yes/no questions, if the answer is no, score 0 for all options.

end, the scores for part X were: CNC, 16; FMS, 14; conventional cell, 0. The part would thus go to a CNC-supported machining cell.

Flexible Machine Tools for High Volumes?

Some of the solutions we've arrived at on the Cummins shop floor seem to violate what has been conventional wisdom ever since Ramchandran Jaikumar's 1986 article "Postindustrial Manufacturing." Jaikumar criticized American piece-part manufacturers for using flexible machine tools to run fairly high volumes. The real virtue of a flexible machine, he observed, is in its ability to handle many different kinds of parts simultaneously, even lot sizes of one. "The U.S. companies used FMSs the wrong way–for high-volume production of a few parts rather than for high-variety production of many parts at low cost per unit."[2]

Cummins is one of those companies running fairly high volumes on flexible machines–especially in hybrid cells. However, it has paid off handsomely for us. In the first 18 months after introduction, the new compression brakes for our engines underwent 14 design changes. On less flexible lines, these changes would have probably taken several years and many millions of dollars to incorporate, which would only have inhibited development of performance-enhancing modifications. Again, the decision to produce this component on an FMS increased cycle time considerably. But it reduced time to market by perhaps two years, reduced annual warranty expenses by an estimated $300,000, and reduced cost to the customer by more than 30%.

None of this is to refute Jaikumar's fundamental insight. We too use flexible machines to run many different low-volume parts, though not necessarily to hit new, customized markets. Flexible machines, we've found, are equally exciting in their consolidation of production lines for obsolescent products. In 1986, we acquired a six-machine FMS to consolidate the production of many of these unpredictable, low-volume parts. It's turned out to be a fairly inexpensive way to keep old customers happy. (The production of water pumps and lubrication system components for many of Cummins's older and specialized industrial engines used to be scattered across several lines of aging machines; setups, changeovers, and downtime accounted for more than 50% of those machines' production hours.)

> New CNC machines, like personal computers, are getting smart enough to stand alone – or to integrate with dedicated lines.

My point is that flexible machines may be important in ways Jaikumar has not specifically addressed: as a hedge against design changes, a push in the race to market, a means to service loyal customers. They are also a saving grace for dedicated machining cells, where part variations, though not really huge, are big enough to make changeover troublesome and where downtime costs more to Cummins than the flexible machine itself.

It's worth noting that during the past four years, stand-alone CNC machines (much like computers) have been getting smarter, cheaper, and easier to program and integrate. Once, buying flexibility meant buying a whole, exotic cell, one in which machines essentially spoke their own language – a kind of machining mainframe. Now stand-alone CNC machines with conversational programming can talk much more easily to their operators, to their environment, even to scheduling systems. They often cost less than dedicated machines (which admittedly are much faster) and can hold tolerances nearly as well.

Flywheels: Proof of the Pudding

It was time to put theory into practice. Flywheel housings had been our biggest headache. Consultants from a Big Eight accounting firm advised sourcing out a considerable portion of this line. We decided to attack the problem instead. We split flywheels into a high-volume flow line and a low-volume cell. Of the original 160 part types machined in the area, only 11, which comprised more than 85% of the volume, remained on the refocused high-volume line. The remaining 149 were then reprocessed into a newly acquired flexible machining cell.

This project, which took about three years and $3 million to complete, has proven to be a magnificent success, as the table "Results of Flywheel Housing Refocus" illustrates. While the initial emphasis was more on improving delivery than on reducing costs, product costs have plummeted more than 40%, resulting in a return on investment of approximately 90%. With these improvements, Cummins has become the lowest cost manufacturer of flywheel housings with a near-perfect delivery record.

Flywheels were a problem also at the engine factory of our joint venture in India. Once again, we had to rationalize the flywheel production area, where high-performance, special-purpose machines are saddled with a wide variety of parts, including flywheel housings and oil pans, supports, brackets, and covers. And compounding the problem were components that had to go to another department because they required intermediate CNC operations.

We divided the area into four cells, each with a well-defined flow: flywheels, flywheel housings, oil pans, and miscellaneous parts. Machines dedicated to individual cells eliminated cross-flow of materials. These measures restored product focus. We then focused volume by integrating a CNC machining center into the flywheel housing line to machine low-volume housings. We subcontracted high-volume housings to small, local manufacturing units with substantially lower overhead.

The benefits of this reorganization were extraordinary: 95% reduction in material handling; 30% reduction in floor space; 50% reduction in WIP inventory; 50% reduction in lead time. We also reck-

Results of Flywheel Housing Refocus

Criterion	*Before*	*After*
Labor efficiency	42%	96%
Hours of output per worker	3.6	7.7
Employees	49	20
Salaried employees	6	2
Throughput time	1.83 days	40 min.
Internal distance traveled	650 feet	256 feet
Inventory reduction		
– raw	6 days	1 day
– WIP	1.4 days	1 hour
– finished	4 days	1 day
On-time delivery	30% on time	100% on time
Product cost	–	43% reduction
Manufacturing scrap	–	99% reduction
Full-time rework employees	3	0
Full-time inspectors	3	Audit only

oned significant but unquantified reductions in direct labor and late deliveries.

In yet another reorganization at our Columbus engine plant, we consolidated the production of many low-volume and obsolescent water and lubrication system components onto a six-machine FMS. These parts were previously scattered across several lines of machines; setups, changeovers, and downtime accounted for more than 50% of production hours. We realized substantial benefits through this consolidation: 50% reduction in direct labor and material handling; 70% reduction in floor space and number of machines; 75% reduction in WIP inventory; 50% reduction in manufacturing scrap; and a reduction in lead time from 42 days to 1 day.

At Cummins, we realized that to face the new manufacturing challenges, we needed nothing short of a total and continuous reorganization of our facilities. We accomplished this through a four-step procedure.

1. Lay out the factory into product-focused cells, routing families or similar products across tightly clustered groups of machines.
2. Once production is running smoothly, reorganize again, dividing the factory conceptually into different classes of production on the basis of volume, design stability, and predictability of demand.
3. Map each product, depending on its production requirements, onto the appropriate class of production machinery. Making parts on the wrong class of equipment–low-volume parts on transfer lines, for example–leads to a loss of flexibility, inefficiencies, and, eventually, uncompetitiveness.
4. Continuously manage the transition of parts from one class of production to another as product life cycles evolve.

There is nothing radical about this proposal, which is basically JIT manufacturing extended beyond the narrow sense of cellular manufacturing and setup reduction. The orchestrated application of these concepts will yield a flexible factory–the real industrial battleground of the next decade.

Reprint 90214

Make or Buy?

The solid corporation acquires suppliers along with technology.

Manufacturing: The New Case for Vertical Integration

by TED KUMPE and PIET T. BOLWIJN

Ever since Robert Hayes and the late William Abernathy exposed the ways we manage our companies into decline, executives in big manufacturing corporations have been struggling with a central strategic problem.[1] It is to find the right balance between investing in vertical integration and encouraging process technology development among suppliers.

The problem for managers is "not their reluctance to take action and make investments," Hayes and Abernathy observed, "but that, when they do so, their action has the unintended result of reinforcing the status quo. In deciding to integrate backward because of apparent short-term rewards, managers often restrict their ability to strike out in innovative directions in the future."

By "innovative directions," Hayes and Abernathy meant strategic approaches that absorb into the production chain the most technologically advanced and cost-effective ways of making components. For any manufacturer, they conceded, backward integration eliminates some purchasing and marketing functions, centralizes overhead, and allows the pooling of R&D and design efforts; where components are, in effect, commodities (ferrous metals or petroleum, for example), backward integration almost certainly boosts profits. Nevertheless, for a technologically active company–one that makes sophisticated consumer electronics or durables–they saw a serious risk. Indeed, Hayes and Abernathy suggested that such a company's best course would be to bid for individual components, not for other businesses.

"[Managers] may suddenly discover that their decision to make rather than buy important parts has locked their companies into an outdated technology." They may find themselves "shut off from the R&D efforts of various independent suppliers by becoming their competitor." Moreover, when a company commits its time and resources to master technology back up the channel of supply, it may be distracted from doing its job well. "Long-term contracts and long-term relationships with suppliers can

1. Robert H. Hayes and William J. Abernathy, "Managing Our Way to Economic Decline," HBR July-August 1980, p. 67.

Ted Kumpe and Piet Bolwijn are senior management consultants with Philips International B.V., Eindhoven, the Netherlands. They are, with Jaap Boorama, Rinus van Breukelen, and Bas Brinkman, authors of the book Flexible Manufacturing: Integrating Technological and Social Innovation *(Elsevier Science Publishers B.V., 1986). Mr. Kumpe is professor of manufacturing organization at Catholic University of Brabant. Mr. Bolwijn is professor of industrial management at the University of Twente.*

"Nice to meet you, Benson. Any friend of free market forces is a friend of mine."

achieve many of the same cost benefits as backward integration without calling into question a company's ability to innovate or respond to innovation."

The further development of computer-driven manufacturing systems during the years since Hayes and Abernathy issued their warning has certainly vindicated a part of their argument—the part that emphasizes the revolutionary benefits of technology. We know a good deal about what it takes to absorb advanced manufacturing systems into the plant—more precisely, how they transform the plant; we know how expensive and taxing automation can be.

General Electric has been working for five years to transform its Appliance Park complex in Louisville, Kentucky (where it builds dishwashers, washers and dryers, refrigerators, and electric ranges) into a highly automated operation. Islands of automation introduced at the dishwasher facility in 1983 have slashed cycle times from five days to several hours, and have helped GE boost market share from 30% to more than 40%. The new refrigerator line, on which $100 million was spent, is automated from one end of the plant to the other. All told, output at Appliance Park has increased while employment has declined from 19,000 to 10,000 over the past decade. GE managers hardly needed the distractions of running components businesses not already their own during this high-stakes transition.

At the same time, Hayes and Abernathy have been proved right about the devolution of important design and R&D functions onto upstream producers. Many smaller suppliers, especially parts makers in the auto industry, have gained skill in computer-aided design and manufacturing systems. Many manufacturers of finished goods have come to depend on innovative suppliers and machine tool vendors to develop robust parts and more efficient manfacturing systems—tasks they used to perform on their own. On the other hand, a good many independent suppliers should now be considered members of the big corporation's family—not owned, but so beholden to the big corporation that their business plans are controlled from above. Is the age of vertical integration over?

We think not. In our view, the solid corporation will continue to view vertical integration as a critical part of manufacturing reform. No doubt, major manufacturers have to learn to get the most from suppliers. But manufacturing reform and backward integration are related in subtle ways to the three stages of production over which the big manufacturers preside. Without integration, technology-based corporations may wind up beggaring upstream components producers—businesses that need the most investment—in order to earn premiums for downstream assembly and distribution operations, businesses that are comparatively flush. This cannot go on indefinitely.

Three Tiers of Production

The advantages of backward integration become clearer when we look closely at the levels of production and at the market forces and advanced technologies shaping each one.

Generally speaking, the production process can be divided into three stages: the assembly stage, the subassembly stage, and the component stage; at each there is a separate tier of factories and businesses. Philips' compact disc players, for example, are produced in plants designed to make finished players from subassemblies, typically, a deck—the chassis into which the motor, laser, and other parts have been integrated—a printed circuit board, and so on. More familiar, perhaps, are subassemblies for cars: engines, gearboxes, or brakes.

In most cases, subassemblies are produced at some distance from assembly plants, either by independent companies—that is, original equipment manufacturer (OEM) suppliers – or by wholly (or partially) owned subsidiaries. Philips owns 400 factories, several of which produce subassemblies such as picture tubes, motors for VCRs or CD players, and decks for VCRs – not only for Philips but also for some of its competitors. The picture tube factories, for instance, deliver roughly 50% of their turnover to non-Philips customers.

At the base of the production pyramid are the components makers; the component stage is to subassembly what subassembly is to assembly. The components of a compact disc player – molded plastics, chips, resistors, and wires – may come from independent or subsidiary businesses, large or small. A car's components are the balance shafts or pistons incorporated into the engine, or the stamped sheet-metal parts that are welded into unpainted "white-bodies." Philips' component factories produce resistors, capacitors, metalware, semiconductors, plastic parts, and other items. These units also act as suppliers for customers both inside and outside the company.

Advanced manufacturing systems affect each tier of production in a different way so that, at each, the cost of failing to invest in process technology is quite different. Each tier competes to serve a distinct market. Each tier will have special problems with respect to production flexibility and quality control.

Consider, first, the changing strategic and manufacturing priorities of assemblers, the big industrial corporations themselves. During the 1950s, in a world economy characterized by scarcity, price was the most important criterion for any consumer. People were happy to have a radio or car; manufacturers were happy to mass-produce hard goods, most of which now seem terribly primitive, in huge batches along moving assembly lines – and with the cheapest labor possible. During the 1970s, following the sharp rise in prosperity, quality became as important for consumers as price. Increasingly, consumers are demanding quality products more nearly tailored to individual needs and tastes – products for young professionals, families, and retirees.

And so corporations like Philips need to be particularly good at manufacturing flexibility, the production of customized products aimed at market niches. To compete effectively in world markets we must differentiate our product lines from the start, without sacrificing quality or price. In the middle of the game we must be flexible enough to increase volume sharply for segments in which demand proves high. Finally, we must redesign our products relentlessly, so they do not outstay their welcome.

Just look at audio/video products, whose commercial life cycles have dropped to a year, in some cases even less. Philips produced about 100 different color television models in 1972. At present, its product line worldwide includes more than 500 different models. As for CD players, Philips puts a new generation on the market almost every year, and each generation offers a jump forward in styling and technology. Philips produced 10 types of players in 1982, 150 today. Manufacturing corporations that do not work to shorten product life cycles in this way will almost certainly be leapfrogged by the competition.

How Assemblers Do Well

And so new marketing pressures and technological advances are causing Philips to rethink its investment priorities. What kind of manufacturing strategy is right for assemblers? In the short run, it is probably not a good idea to invest heavily in advanced flexible manufacturing systems – not yet. Managers still cannot buy robots that are both flexible and inexpensive enough to handle major design changes in end products, whose life cycles are inherently short. It is more sensible for assemblers to divide a plant into focused factories within factories; more cost-effective to pay motivated people and pay for more or less dedicated machines – machines that are flexible enough to handle design variations in one generation of compact discs or VCRs and that can then be retooled or even discarded when the new generation is born.[2] In the Wetzlar factory, Philips builds car radios with a modular assembly system in which each standard module performs only a limited number of assembly tasks. Resetting the system for a new generation of radios has thus far proved a simple operation.

The key flexibility for assemblers in the longer run is the setup time required to move from one generation of products to the next. When delivering products quickly and reacting to changing market demands by shortening product cycles, economies of scale are, of course, not worth pursuing. More important efficiencies can be gained from reductions in work-in-progress inventories, stock inventories, throughput time, and transportation costs.

These considerations make it more attractive than ever to set up assembly factories close to end markets. They make the introduction of just-in-time operations necessary. As a matter of fact, industrial corporations add value (and realize high profit margins) less by refining the process of assembly than

2. For an advocacy of this viewpoint, see Richard J. Schonberger, "Frugal Manufacturing," HBR September-October 1987, p. 95.

by marketing aggressively, distributing shrewdly, and backing up their retail customers with fast delivery, satisfactory service, and appealing design.

There are exceptions, to be sure. Where the market is stable (and competition from the Japanese unlikely), a big manufacturer may well justify a long-term investment in automation at assembly. GE's refrigerator market is a case in point. Still, most hard-goods producers do not have the luxury of planning this far ahead. For most of us, the secret of assembly is flexibility, and the secret of flexible assembly is the organization of rather simple machines with people–those most complex and flexible of beings. When you think about it, Volvo's advances in the work team concept may have done more to teach big manufacturers about our "factories of the future" than GE's advances with automation.[3]

The Philips vacuum cleaner plant at Hoogeveen, Holland recently went through a reorganization to simplify product flow, make the managerial organization leaner and flatter, and institute fast information feedback loops including specialists from development, engineering, production, and marketing. Ideally, the management of assembly will be visible and vigorous. Reporting relationships will be cooperative rather than authoritarian. Assembly systems will ensure rapid communications and feedback of information throughout the organization. Simple, clear-cut, product-oriented organization will break down barriers among design engineers, manufacturing executives, and marketing specialists. Flexibility and bureaucracy do not mix.

None of this is to deny that robotics will likely advance us to the point where assemblers–those that have perfected flow operations–can replace people. Then, automation will mean even more consistent quality at assembly and fewer interruptions on the line; computer-driven refinements in measuring and calibration are already reducing the time operators need to stop the line to reset machines or log in quality control data. Nevertheless, important advances in quality at assembly nowadays are coming not from the architecture of chips so much as from the organization of people, from people-integrated manufacturing, not from computer-integrated manufacturing.

Subassemblies, Components, Quality

The question of quality, of how major manufacturers build it in at assembly, brings us to new and decisive relations among the three tiers of production. It also brings us to the heart of the matter. That a motivated and well-organized work force is important for assembling quality products is hardly arguable. The *most* important advances in quality, however, are coming not from anything assemblers do in their factories but from what subassemblers do in theirs.

During the past five years, hard-goods industries have achieved a striking reduction in the number of parts and subassemblies that have to be fit together to make an end product. It is this reduction that, more than anything else, accounts for increasing product quality. Philips, the world's largest manufacturer of compact disc players, has reduced the number of subassemblies by 75% over the last five years. The most prescient manufacturing corporations have been working with OEM suppliers to design subassemblies that are more manufacturable–that include more functions but fewer components and are designed to fit together in more predictable ways. Fewer parts mean fewer opportunities for error.

Perhaps the best known of these product programs is the IBM Proprinter, most of whose resilient molded parts snap together and require perhaps one-third the number of assembly operations as earlier model printers. The Proprinter's subassemblies are so manufacturable that robots can put them together, while doing the job manually takes only about three minutes. (Revealingly, after building the units for machine assembly, IBM transferred assembly operations from a robotized line in Charlotte, North Carolina to direct-labor lines at its typewriter facility in Lexington, Kentucky.)

Of course, getting the number of subassemblies down has been a terribly expensive proposition. Early on, IBM committed manufacturing engineers to develop subassemblies for the Proprinter, including investments in robotized injection molding machines and fused plastic/metal components. Subassembly businesses generally soak up a comparatively higher investment in automation and R&D than assembly businesses. Philips invested more than $250 million in a flat, square picture tube for a new generation of color televisions (CTVs).

It should come as no surprise, then, that the logic working on relations between assemblers and subassemblers works also on relations between components producers and subassemblers–only more so. Technology improvements and design innovations that have increased the complexity of electrical products and vehicles have at the same time greatly reduced the complexity of assembling them from components and subassemblies. New generations of semiconductors, integrated plastic and metal parts,

3. We explore this development in "Toward the Factory of the Future," *McKinsey Quarterly*, Spring 1986, p. 40.

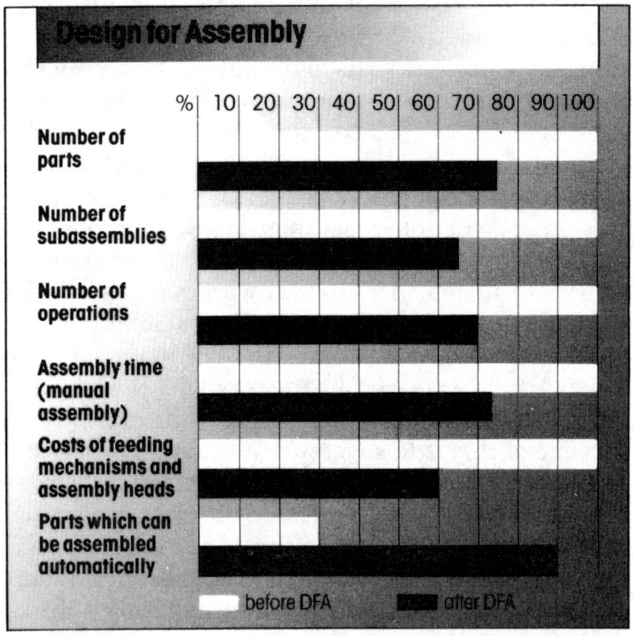

We have studied 29 products including compact disc players, video cameras, telephones, and portable irons, from 1984 to 1986. In all, there have been striking reductions in the number of components and assembly operations, and corresponding reductions in cost.

ceramics, surface-mounted devices, switches–all these have led to striking efficiencies.

Since 1984, Philips has reduced the number of parts in its compact disc player by 75%. It has simplified VCRs by a parts reduction of 55%, tuners by 45%, and amplifiers by 40%. And the trend has not been restricted to the makers of audio/visual products. Swatch has reduced the number of components in its watches from 150 to 51. Less dramatically, though more significantly perhaps, Ford has reduced the number of components in its average car from 30,000 to 22,000. Similar changes have been made in microwave ovens, personal computers, office copiers, and air compressors.

These reductions have meant improved productivity rates at assembly. They have also meant, clearly, decreased value-added from assembly operations as a whole. For end products, even those that depend almost entirely on direct labor to fit their parts together, final assembly constitutes only a small part of the total labor content. There has been a strong shift in the distribution of value-added toward the components stage. In 1969, assembly of a CTV took 12½ hours; today the process takes approximately 60 minutes, and the elapsed time will go down considerably within the next few years.

At Philips, the cost structure of consumer electronic products–CDs, CTVs, PCs, or word processors–typically consists of 70% material, 5% direct labor, and 25% other costs like indirect labor, buildings, and interest. Today there is more added value in building the picture tube than in assembling the whole CTV. A corresponding shift can be seen in investment: in 1986, components amounted to 13% of the $26 billion turnover, but 30% of all company investment went to the components area. In the future, components will take an even bigger slice of the investment pie.

Incidentally, one can readily see why opportunities for introducing advanced technology at the components stage tends to be greater than at other stages. As we said, flexible manufacturing equipment that is big enough and smart enough to handle major changes in assembly routines is not likely to be cheap and friendly enough. But equipment that allows makers of smaller machined or molded components to be truly flexible are much cheaper, much friendlier, and already up and running. Ramchandran Jaikumar writes acutely that FMS technology allows parts makers–the healthy ones–to produce virtually any component at very high quality and at a cost equal to what used to be expected only from mass production.[4]

And so where factories producing subassemblies are tending toward CAD/CAM and a high degree of flexible automation, the trend among component factories is from high-precision mechanics to ultra-high-precision mechanics, from micron technology to submicron technology, from plastic or metal parts to integrated plastic/metal parts. Tolerances in these factories have become very exacting. Costs in R&D are unprecedented.

Suppliers Get Bigger, Not Richer

Given the enormous R&D investment required to manufacture components competitively, it seems obvious that components makers would tend to get bigger and bigger–that is, fewer and fewer–and would have to enjoy relatively high profit margins. Jaikumar argues that because of the new FMS technology, size is no longer a barrier to entry. But there are other factors which put small components makers at a distinct disadvantage.

To understand these, it is useful to divide the components industry into two areas: microelectronics and mechanical parts. The microelectronics area presents a clear case for bigness. Philips' newest chip factory in Nijmegen, Holland cost more than $250

4. Ramchandran Jaikumar, "Postindustrial Manufacturing," HBR November-December 1986, p. 69.

million. The turnover necessary to recover such an investment profitably is correspondingly high. And developing new components in this area may also mean developing new materials and processes. Product and process cannot be viewed separately; Philips and Siemens together spent about $1 billion mastering submicron technology.

The traditional Swiss watch industry was not vertically integrated. It consisted of watchmakers that designed and assembled the components of a great many small independent suppliers. When the Japanese watch industry gained a large portion of the world watch market by supplying quartz watches, the Swiss were taken aback. The quartz technology was totally new to Swiss suppliers, and many of them could not and would not adopt it. Swatch, the symbol of the Swiss recovery, is the result of a concerted effort involving all steps of the production chain from components to final assembly.

But high-tech components are found in the mechanical area too, which is why German automaker Daimler-Benz took over AEG (a $3.5 billion company). Daimler-Benz also absorbed Dornier ($750 million) for its know-how in lightweight carbon fibers and synthetic alloys, and MTU ($1 billion) for its research and manufacturing capacity in ceramics. Large components suppliers, having wider technology bases, are in a better position to introduce new technology in any part of their businesses than are small suppliers.

And so with respect to size, the picture is as one would expect. Components manufacturers – the ones that have survived – have grown enormously over the last ten years, a trend that is clear whether one thinks of electronics components makers such as Emerson Electric, or comparatively small automotive components makers such as Simpson Industries. With respect to the growth of margins, however, the picture is surprisingly bleak. Profits are on the whole much higher for the makers and sellers of end products than for the makers and sellers of components. Big profits come from selling computers or automobiles, not from selling chips or shafts.

The reasons for this pattern are all too clear. Margins on general-purpose components – switches, wires, resistors, bolts – are lower than on customized components. And for many of these low-tech components, high-tech replacements with improved price/performance ratios often appear – all of which lead to dumping and the further shrinking of margins. True, the makers of standard components profit from large-scale production. But most of the benefits of scale are mitigated by brutal competition.

As for the costs of customized components, *their* development can be amortized only over a long time – at least three years in the auto industry. Yet the success of the end product into which customized components are designed is almost entirely out of the components makers' hands. Philips or Ford will introduce a range of products over the next five years; consumers will not accept them all. But components makers will have to bid to make the things that go into them all and, in many cases, produce prototypes along with their bids. These producers cannot expect to recover their costs for all the components they design – another reason, by the way, why the versatility that comes with bigness is an advantage.

None of this is to suggest that major manufacturing designers aim to build end products mainly from customized components. In fact, the opposite is true. Our best design strategy is to build as much as possible from standard parts but to develop a proprietary advantage in certain critical components – chips, decks, printed circuits, and so forth. Of course this strategy only adds to the components makers' difficulties. It tends to eliminate customization, hence opportunities to make high-margin items.

To be blunt, the typical big end-product manufacturer contributes a disproportionately small share of the value-added, while the typical components manufacturer reaps a disproportionately small share of the profit. Heads, the major manufacturer wins; tails, the components maker loses. All of which, obviously, brings us back to the subject of vertical integration.

Since most of the technology content of end products is in components, where earnings are shallow, major producers of technologically sophisticated goods will have to compete more aggressively for world leadership in component development and production – something the Japanese companies have been doing all along. Look at the 1986 ranking of the top integrated-circuit manufacturers by sales:

1. NEC
2. Hitachi
3. Toshiba
4. Motorola
5. Texas Instruments
6. Philips
7. Fujitsu
8. Matsushita
9. Mitsubishi
10. Intel

The producers of end products that need microprocessors are slowly taking over this business; the list does not even include companies such as IBM (the world's largest producer of chips) and Hewlett-Packard, both of which make chips solely for their own use.

True enough, big manufacturers may want to take over certain components operations in any event, if only to make it easier to coordinate design functions. Our point is that, regardless of design considerations, the financial pressures will ultimately force them to take over more components makers. Large, multinational electronics companies like Philips are especially driven in this direction. Only within the integrated company will there be an assured flow of advanced components to end product manufacturers and a flow of capital back to components manufacturers. Only through vertical integration is the circle firmly closed.

Manufacturers that do not invest in vertical integration may, for a while, enjoy some cash advantages. Eventually, however, they will drive components producers out of business or, more likely, find themselves buying components from their own competitors. Philips took over Signetics in the 1970s to avoid such competition. Recently, Philips built its own factory for producing liquid crystal display (LCD) screens.

So the question is one of business goals. If you intend to maximize returns on investments, you may well decide to sell off suppliers. If, however, you strive for world market leadership, you will strive for vertical integration.

Will Competitors Be Your Suppliers?

In fairness to Hayes and Abernathy, conditions in electronics industries are somewhat different from conditions in, say, the auto industry. Auto suppliers may be so much a part of the family that major car companies will unilaterally guarantee the flow of capital back to high-quality suppliers either to keep them solvent or to underwrite promising research ventures. The makers of machine-vision quality control equipment springs to mind.

And yet automobile suppliers in Michigan, as everywhere else, are hardly thriving. The refrain one hears again and again is that stockholders are losing patience with razor-thin margins. If suppliers could get together and build their own car, it would be another matter. But that, for components makers, is always another matter.

Viewed in this light, the growing strength of Far Eastern companies (especially Japanese) in the components industry constitutes a definite threat for Western high-volume manufacturers. In some areas, Far Eastern companies already hold an almost monopolistic position. Even in the important area of semiconductors, a traditional U.S. stronghold, we have seen the Japanese making major inroads.

The unabating pressure on prices will force an unabating search for components with better price performance ratios. Future developments in the television industry, for example, will take place in integrated circuits, surface mount devices, and picture tubes. A producer of television sets not sufficiently involved in the research in these areas will be in a very weak competitive position. (Philips is researching new generations of high-definition CRT tubes, wide-screen tubes, new processing chips, projection TVs, and LCDs.) Many improvements in the car industry will come out of electronics and new materials like ceramics. But will Western industries on the whole become dependent on integrated Japanese companies for critical components?

All in all, the tendency of many Western companies toward less vertical integration seems increasingly unrealistic. Flexibility downstream and innovation upstream are critical to manufacturing renewal. And this means that big companies will have to integrate their activities along the business chain and carefully plan to transfer capital from one link to another. Undoubtedly, manufacturers face a new world. The solid corporation of the future will, for new reasons, resemble the solid corporations of the past.

Reprint 88206

Small-company focus and *large-company efficiency?*
Try a "VAP."

Beyond Vertical Integration – the Rise of the Value-Adding Partnership

by RUSSELL JOHNSTON and
PAUL R. LAWRENCE

For decades large, vertically integrated companies have reaped the benefits of their size, growing stronger with every competitor they eliminated or engulfed. But the elephants aren't grazing so freely anymore. Another beast has been nibbling at the herbage, and its presence is beginning to be felt.

That beast is the "value-adding partnership" – a set of independent companies that work closely together to manage the flow of goods and services along the entire value-added chain. It is an organizational form much like the putting-out system of the early industrial revolution, whereby manufacturing was done in cottages and coordinated by a merchant-manufacturer who supplied the raw materials and sold the final product. But the value-adding partnership, or VAP, is not an anachronism. It is a product of its time, and its time may well have come.

Most historians agree that the development of cheap, centralized power and efficient but costly production machinery tipped the competitive advantage toward large companies that could achieve economies of scale. Today, low-cost computing and communication seem to be tipping the competitive advantage back toward partnerships of smaller companies, each of which performs one part of the value-added chain and coordinates its activities with the rest of the chain.

VAPs are not, however, necessarily technology driven. They may emerge as the result of computerized links between companies or they may exist before the technical links have been made. In all cases, they depend largely on the attitudes and practices of the participating managers. Computers simply make it easier to communicate, share information, and respond quickly to shifts in demand. They facilitate VAPs but alone don't create them.

To better understand what a value-adding partnership is and how it works, let's look at some that are doing especially well.[1]

McKesson Corporation, the $6.67 billion distributor of drugs, health care products, and other consumer goods, is among the most successful. The business press has often cited McKesson for its innovative use of information technology to improve customer service and cut order-entry costs. But McKesson's story is much richer than most people know. Once a conventional wholesale distributor squeezed by vertically integrated chain stores, McKesson has transformed itself into the hub of a large value-adding partnership that can more than hold its own against the chains.

McKesson's evolution to a VAP was triggered by fierce competition from large drugstore chains,

Russell Johnston is an associate professor of management at Boston University, where he studies the effect of increasing access to information on management processes and structures. Paul R. Lawrence is the Wallace Brett Donham Professor of Organizational Behavior at the Harvard Business School and is the author of numerous books and articles.

which were eating into the business of the independent stores McKesson serviced. McKesson realized that if the independents died, it would soon follow suit. To protect their business, McKesson's managers began to look for ways to help customers.

Their search focused on a rudimentary order-entry system at one of McKesson's warehouses. In the early stages, the system included data-collection devices, powered by car batteries, that were wheeled around customers' stores in shopping carts. The system dramatically cut the costs of processing orders by expediting the steps of checking inventory, calling in an order, manually recording the order, and eventually packing and shipping it. McKesson soon discovered that the system could also specify how to pack orders so that they coincided with the arrangement of customers' shelves. Doing so made restocking more efficient.

These successful uses of information technology spurred the search for others. McKesson managers soon realized they could use the computer to manipulate data to help customers set prices and design store layouts to maximize the profits of each particular store. They also began using it to perform accounting services, such as producing balance sheets and income statements. And they discovered that the system could be used to warn consumers of potentially harmful drug combinations by tracking prescription histories.

McKesson thus offered the independent drugstores many advantages of computerized systems that no one store could afford by itself. The drugstores were able to offer their customers better prices, a more targeted product mix, and better service, all of which helped them stand up against the chains. Still, the drugstores maintained their autonomy, so they could be responsive to the needs of the local area and form lasting ties with the community. This actually gave them an advantage over the chain stores, whose managers had to answer to headquarters and could be transferred from one location to another.

McKesson, of course, benefited from the independents' good health. The user fees covered the cost of service development plus provided a return on the investment. Since the system was introduced in 1976, sales to pharmacies have soared from $900 million to over $5 billion. And the more efficient ordering systems allowed the company to reduce its warehouses from 130 to 54, eliminate 500 clerical jobs devoted to taking telephone orders, strengthen its customer base from 20,000 customers averaging $4,000 a month in orders to 15,000 customers averaging $12,000 to $15,000 a month, and reduce the average number of shipments per customer from two per day to two per week while lowering its own and customers' inventory costs.

The close and productive link with customers wasn't good enough, however, to satisfy McKesson's imagination. The company recognized that the up-to-date information on sales had immense value to product managers of consumer goods manufacturers and proceeded to sell it to its own suppliers. Suppliers used it to make more timely shipments to McKesson in much the same way as McKesson had done with the drugstores. Computer-to-computer ordering from suppliers permitted McKesson to cut its staff of buyers from 140 to 12. Meanwhile, suppliers could schedule production more efficiently and streamline their inventories.

> In a VAP, companies form close and lasting ties with others in the value-added chain.

Another McKesson innovation was to use the computer system to help process insurance claim applications for prescription reimbursement. This strengthened the ties among insurance companies, consumers, and drugstores by speeding payments and smoothing administrative hassles. McKesson's total network thus includes manufacturer, distributor, retailer, consumer, and third-party insurance supplier.

What makes McKesson so powerful—and what makes it a VAP—is the understanding that each player in the value-added chain has a stake in the others' success. McKesson managers see the entire VAP—not just one part of it—as one competitive unit. It was this awareness that allowed McKesson's managers to look for opportunities beyond their own corporate boundaries. They looked for ways the resources at one part of the value-added chain could be used in another. And their efforts to be competitive went beyond cost cutting. Many companies focus on trimming costs to increase profits, and they consider opportunities only within the unit defined by ownership. McKesson also looks for ways to add value by creating new services.

1. For other discussions of the new organizational forms, see Raymond Miles and Charles Snow, "Network Organizations: New Concepts for New Forms," *California Management Review*, Spring 1987, p. 62; Robert G. Eccles, "The Quasifirm in the Construction Industry," *Journal of Economic Behavior and Organization*, December 1981, p. 335; Calvin Pava, "Managing the New Information Technology: Design or Default?" in *HRM Trends and Challenges*, ed. Richard E. Walton and Paul R. Lawrence, (Boston: Harvard Business School Press, 1985); and Andrea Larson, "Networks as Organizations," unpublished manuscript, 1987.

What's a Value-Added Chain?

The term *value-added chain* comes from the field of microeconomics, where it is used to describe the various steps a good or service goes through from raw material to final consumption. Economics has traditionally conceived of transactions between steps in the chain as being arm's-length relationships or hierarchies of common ownership. Value-adding partnerships are an alternative to those two types of relationships. Usually, the partnerships first develop between organizations that perform adjacent steps in the chain.

A value-added chain for packaged foods might look like this:

farmer → broker → basic processor → packaged goods producer → distributor → retailer → consumer

This ability to see beyond the corporate boundaries has another important advantage. It permits recognition of serious threats that lie elsewhere along the value-added chain. Because McKesson knows its own fate depends on that of its suppliers and customers, the company monitors competitive dynamics throughout the chain and tries to fix weaknesses wherever they occur. When all the partners are strong, the entire value-added chain can stand up to the toughest of competitors, integrated or not.

The McKesson partnership is so successful that others in the pharmacy distribution business have emulated it or withdrawn entirely. But most have missed the point. It is easy to make the mistake of thinking that McKesson's network is nothing more than a computer system with terminals in someone else's building. The wires and processors are not what make McKesson successful. True, the McKesson VAP grew out of the company's computer system, but information technology did not create the VAP. Rather, it was the managers who understood the relationships along the entire value-added chain and the need for each link in the chain to be as strong as possible. Information technology is not even a necessary ingredient in a VAP, as the next example demonstrates.

The textile industry of central Italy comprises many successful VAPs, which have evolved very differently from McKesson.[2] Over the past 20 years, 15,000 to 20,000 smaller companies have replaced all but one of the large, vertically integrated textile mills of the Prato area. By 1982, these companies were employing 70,000 people and exporting about $1.5 billion worth of products. The industry's disintegration may have begun partly to avoid labor legislation, but the new structure has allowed the industry to thrive for more basic reasons.

The Italian story really begins in the early 1970s, when Massimo Menichetti took over a large, integrated textile mill from his father.[3] At that time the company's future–indeed that of the whole Italian textile industry–looked bleak. Labor costs were soaring throughout Italy, and foreign competition was intensifying. Furthermore, a trend toward greater product variety meant that companies had to be able to create new designs quickly and efficiently, shifting production from one product to another without wasting time or materials. Innovation and flexibility had become critical to survival. Increasingly squeezed between rising production costs and falling market prices, Menichetti's mill had been losing money for several years.

Menichetti believed that the company had become too big and bureaucratic to adapt to the new competitive demands, so he proceeded to break the company into eight independent organizations. He arranged to sell between 30% and 50% of the stock in those companies to key employees, who would make the purchases with company profits–thus enabling them to become part owners without putting up any of their own money. The ownership transfer was to be gradual, over the course of three years. By the end of that period, the new enterprises would have to make half their sales to outside companies–to avoid a slip into

2. This description draws heavily on Michael J. Piore and Charles F. Sabel, *The Second Industrial Divide* (New York: Basic Books, 1984) and on Gianni Lorenzoni, *Una Politica Innovative* (Milan: Etas Libri, 1979).
3. The facts about Massimo Menichetti are excerpted from the HBS case, Massimo Menichetti (B) 686-135, revised October 1986, prepared by Ramchandran Jaikumar.

complacency. To demonstrate his resolve to play only an advisory role and keep out of operations, Menichetti also started a marketing company in New York. He stipulated that it could represent no more than 30% of the production volume of the Menichetti group.

Within three years, the dismantling of the Menichetti mill was complete and business was being conducted very differently. Since then, other integrated mills have patterned themselves after the Menichetti VAP. Small companies with cooperative relationships are now spread through the entire Prato area of central Italy.

> VAPs like Menichetti are common in the Italian textile industry.

Formerly, in each large mill, one group of managers oversaw the entire process, from assessing the market to designing fabric to supervising every detail of production. Now, small groups—sometimes a family—take total responsibility for their part in the process. Each shop has certain special skills. One may be particularly good at producing high-quality knits for dresses; another may be expert at mixing colors. Work is contracted out to whichever shop can meet the market's needs at the time. Each, therefore, has great incentive to stay in touch with fashion trends and environmental changes and to be ready to react quickly. Otherwise, it would lose business to other producers and might even go out of business.

At the center of each set of small companies is an independent master broker, or *impannatore*. In the Menichetti VAP, Massimo Menichetti himself plays this role. The *impannatore* manage the relationships among the various shops. They are facilitators and problem solvers who carry information from one place along the value-added chain to wherever it will be most useful. They get involved in all aspects of the textile business: raw materials purchases, fabric design, production contracts, transportation, and sales. They look at the weaver's samples for next year and if they think they will sell, take them to customers all over the world. If the market objects to the weaver's price, the *impannatore* may help the weaver find ways to trim costs. They also negotiate with raw materials suppliers and transportation providers.

Being close to the customer, the *impannatore* were the first to realize that market changes required increased innovation and flexibility. They took the lesson to heart and, more importantly, carried the word back to the small manufacturers. To avoid losing business because of an inability to react fast enough, the production shops have adopted the latest textile machinery, including numerically controlled looms. Whole chains, not just individual players, adapted quickly to the market information, and all have benefited.

Realizing that their partners must also be financially sound, efficient, and marketwise if they themselves are to be competitive, the players in the Italian textile VAPs are eager to share information and cooperate. In recent years, they have developed computer systems that rush information from partner to partner. The technology enhances coordination and boosts the speed and quality of responses to the market. The computer systems enhance the VAPs, but again, do not create them.

Of course, sharing information is very different from sharing rewards. In a VAP, as in any other industrial or organizational structure, innovation and adaptation must be rewarded if they are to be encouraged. In the Prato area, the *impannatore* can ensure that the rewards are shared appropriately by influencing prices and channeling work only to cooperative members. They can, for instance, withhold work from a shop that is trying to drive out otherwise successful competitors through predatory pricing.

The Prato mammoths gave way quietly and gracefully to the extraordinarily successful VAPs. Systematic, close coordination is now the rule, not the exception. In fact, the ties exist not only vertically, with suppliers and customers, but also horizontally, with what would usually be considered direct competitors. A weaver that guesses wrong one season might well receive overflow orders from a competitor that guessed right. They both understand that next year their roles may be reversed. And they know that if they help each other through tough times, they can avoid building overcapacity that could eventually hurt them all. Computer networks have been extended to interconnect the Italian VAPs, so when one VAP cannot deliver, another can be called on right away.

After only five years, all Menichetti's productive units had over 90% utilization of their machines. Both labor and machine productivity had increased. New machines had been added, increasing capacity by 25%. Product variety was increased in each of the eight units from an average of 600 to 6,000 different yarns. Average in-process and finished-goods inventory dropped from four months to 15 days. What works for Menichetti works for the Italian textile industry as a whole. From 1970 to 1982, Prato production of textiles more than doubled, while that in the rest of Europe declined steeply.

Other VAPs are alive and well and show that Prato and McKesson are not flukes. The construction industry is a third example. It has operated like a value-adding partnership since the time of the Roman Empire. General contractors subcontract almost all the work on a construction job, soliciting bids from a selected set of subcontractors they trust and making contracts with "partners" who offer reasonable prices – not always the lowest bid.

Japanese trading companies are venerable VAPs that are even more extensive than those in the construction industry. They arrange for the buying and selling of goods at every step of the value-added chain, from mines to household consumers, across several continents. They never get involved in operations. Although some Japanese companies are now choosing to develop their own brand images and find their own way to foreign markets, trading companies remain central to Japan's economic success.

Japanese auto companies also operate as VAPs. Toyota, for example, directly produces only 20% or so of the value of its cars, while GM and Ford produce 70% and 50% respectively. Chrysler's comeback was due in part to the creation of a VAP with its suppliers, distributors, and union. It produces only around 30% of the value of the cars it sells. Many industry observers attribute Ford's recent gains on GM to Ford's aggressive moves to form partnerships with suppliers.

In the past 30 years, book publishing has evolved toward a VAP. The leading competitors have taken turns divesting various operations that were formerly vertically integrated. The printing function was one of the first to be farmed out, followed by graphics and artwork. Now the usual core function of publishing is brokerage and marketing.

The movie industry has been moving in a somewhat parallel way. Full-blown movie studios that hold exclusive long-term contracts with actors and directors, have a staff of full-time composers and scriptwriters, and own and operate fully equipped production lots are a thing of the past. Now the studios act like brokers who negotiate a set of contracts for a single film production. Old-fashioned studios have been unable to compete.

"My mother sent down some chicken soup to see if that will help."

At least in theory, whenever a nonintegrated company deals with another company that performs the next phase of the value-added chain, both stand to benefit from the other's success. But usually, such companies hold each other at arm's length and struggle to keep any economic gains to themselves. In fact, organizations often try to weaken a supplier or customer to ensure their own control of profits. This is understandable, given that the widely followed competitive model suggests that companies will lose bargaining power – and therefore the ability to control profits – as suppliers or customers gain strength.

The relationship between companies connected only by free-market business transactions and guided by such a model of competitiveness is often guarded, if not antagonistic, and rooted in fear that the other will become a competitor or engage in some other opportunistic behavior. Naturally, such companies tend to share as little information as possible, and consequently managers often lack knowledge of the activities elsewhere along the value-added chain. If a company perceives a trading partner as an adversary, it may ship shoddy materials, squeeze margins, delay payments, pirate employees, steal ideas, start price wars, or corner a critical resource – all practices that reveal a lack of concern for the supplier's or customer's well-being.

The conventional solution for ending such destructive games and for controlling resources is vertical integration. When organizations along the value chain are under one management, it is presumed that they can coordinate their activities and work toward a common purpose. And, of course, they can often realize economies of scale.

But vertical integration has its weaknesses. In the process of exploiting their distinctive competences, many large, integrated companies emphasize one competitive dimension. In an integrated company, such focus can actually be a liability, because the strong culture that supports that focus makes it hard to perform tasks that require distinctly different orientations and values. A business that emphasizes low cost, for instance, may run its factories well, but its R&D, design, or marketing functions may have trouble innovating. In a chemical company dominated by commodity production, the culture may inhibit specialty operations. The packaging division of a large paper manufacturer that emphasizes mass production may have trouble responding to the market as an independent competitor.

Perhaps the best example of this problem is in manufacturing. Many manufacturing companies that have invested heavily in flexible manufacturing systems in recent years have had trouble making the new technology achieve its potential. The culture and practices that support long production runs of standardized parts don't fit the new emphasis on wide product variety.

The problem of focus applies to horizontally integrated companies as well. A manufacturer of automobile parts is unlikely to be equally successful at making other products. Although the similarities may be many, whatever differences exist are likely to keep one or the other lines from doing as well as it could if it were the company's sole product. And in many companies, large size itself creates a certain complexity that inhibits communication, innovation, and flexibility.

In a VAP, each small operating company focuses on doing just one step of the value-added chain. Therefore, each unit can tailor all aspects of the organization to this single task. Personnel, plant and equipment, compensation schemes, career tracks, accounting systems, and management styles–all vary depending on the work to be done. The drugstores in the McKesson VAP can attend to their customers' needs and let someone else concentrate on getting the products on the shelf at the right time. In the Prato area, the small companies that produce fabric strive for low cost, coupled with flexibility; those who design the fabric emphasize innovation and creativity.

This sense of focus translates into low overhead, lean staff, and few middle managers. Decisions are made and executed quickly, so response time is short. Creative ideas are less likely to be suppressed, and more employees are exposed to the demands of the market. The fact that each company in a VAP is free to be different from the others creates a diversity that can be the seedbed of innovation. And marketing orientation becomes not an edict nor a difficult task. It follows naturally from the free flow of information throughout the value-added chain to so many of the people who actually do the work.

At the same time, value-adding partnerships have some of the advantages of vertically integrated companies. Managers in a VAP take an interest in the success of other companies in the value-added chain. Their partnership orientation means they work toward the common goal of making the whole VAP competitive. They have command of facts about the market and empathy for the other organizations they deal with. Because information is shared throughout the chain, they know a lot about the competition. And they coordinate their activities with those of their trading partners.

VAPs can also secure the benefits of economies of scale by sharing such things as purchasing services, warehouses, research and development centers, and of course information. McKesson's partners share access to the computer system. Partners in the Menichetti VAP are so congenial that they are housed under one roof; lines on the floor mark where one ends and the other begins. And that VAP has a cooperative transport system.

The power of the VAP is undeniable. To a great extent, VAPs have the best of both worlds: the coordination and scale associated with large companies and the flexibility, creativity, and low overhead usually found in small companies. VAPs share knowledge and insight but aren't burdened with guidelines from a distant headquarters. They don't have long forms to fill out and weekly reports to render. They can act promptly, without having to consult a thick manual of standard operating procedures. In an increasing number of industries, they are proving to be fiercely competitive against both large companies and small independents.

> **Is the VAP here to stay, or will it fall prey to power plays?**

Indeed, the spate of failed mergers and subsequent divestitures and spin-offs, what some people call downsizing, demonstrates that conglomerates and vertically integrated corporations are not always the most competitive organizational forms. The largest organizations in the United States seem to be losing their footing. Employment at the approximately 800 companies ranked in the *Forbes* "500" on sales, profits, assets, or market value declined from more than 23 million in 1979 to 20.6 million in 1986. Average employment at the companies declined from nearly 29,000 to just over 26,000 during the same period. These numbers fell despite a significant rise in total

New Technological Tools Help Create VAPs

Tools	*Implications*
Minicomputers and PCs; user friendly languages; inexpensive general-purpose software packages	Drastically improves the economics of small scale, providing wider access to information power to include the *smallest* organizations and the *lowest* organizational levels.
Data standards; bar codes	Enable rapid, inexpensive, accurate capture and use of information in electronic form, lowering transaction costs *between* organizations.
Information networking capability	Permits instantaneous sharing of information *between* organizations with shared interests–thus increasing speed and economy of coordinated response to market changes.
Computer-aided design	Improves speed and economy of response to customer needs by improving coordination *between* organizations in design functions.
Computer-aided manufacturing	Permits use of just-in-time practices *between* organizations.

U.S. employment, not to mention the acquisition programs many of these large companies pursued.

Small size alone is not the answer. Many small companies that have open-market relationships with other businesses survive only at the whim of a larger competitor, customer, or supplier that could readily drive it out of business or acquire it if margins become attractive enough. Always constrained by fierce competitiveness and trading partners that know no loyalty, they have little freedom to make financial and operating decisions that are best in the long run.

The delicate issue of control raises questions about the viability of a VAP over time. Let's not forget how creative businesspeople can be. They can invent dozens of ways to take advantage of each other. What prevents them from playing destructive games with their VAP partners, who are, after all, potential competitors? What prevents hostile takeovers? In short, what is to stop a VAP from devolving into anarchy or back to a vertically integrated giant?

For a VAP to exist, its partners must adopt and adhere to a set of ground rules that generates trustworthy transactions. The sense of partnership must become an enforceable reality, despite the many uncertainties and opportunities for playing games. Advice on the best way to do this comes not only from the examples of successful VAPs but also from economists and political scientists who have experimented with the "prisoner's dilemma." The prisoner's dilemma is a game in which two "prisoners" are separated. Each has the option of either squealing on the other, thereby getting more lenient treatment for himself, or remaining silent, thereby saving both himself and his partner in crime. Of course, if one prisoner remains silent but the partner squeals, the silent person will suffer.

When the games are repeated over and over again, the strategy proving most beneficial is "tit for tat." That is, those players who cooperate on the first round and thereafter do whatever the other player did on the previous move are more successful. Those who don't catch on get eliminated. Robert Axelrod has summarized the extensive studies of the prisoner's dilemma in his 1984 book, *The Evolution of Cooperation*. His advice is particularly relevant to businesspeople in a VAP: (1) don't be the first to play games, (2) reciprocate with both cooperation and lack of it, (3) don't be too greedy, and (4) don't be too clever and try to outsmart your partner.

Studies of existing VAPs are far from conclusive, but early indications are that VAPs follow Axelrod's advice intuitively. They are thus very different from the theoretically perfect markets of economic theory, in which bidders balance supply and demand around price and caveat emptor is the guiding principle.

Each company in a VAP cultivates relationships with only a few (from two to six) suppliers of critical items and customers. Having too many partners means few repeat transactions and no time for close relationships to develop. At the same time, partners avoid becoming overdependent on one relationship. A company can keep potential partners "on reserve" through occasional transactions so its welfare won't be harmed if a regular player fails to cooperate.

If partners are to help one another, VAPs must have ways of sharing information. If a partner's costs are creeping out of line, others must know so they can explore ways of helping with cost controls. Technological developments are making it easier for companies to exchange information (see the accompanying list of computer tools and their implications). But also important, successful VAPs must be able to punish partners for acts of opportunism and gaming.

In the Prato textile area, late delivery sometimes calls for withholding of new orders until the problem is rectified. And the construction industry has invented many ways to cope with changes in job specifications and raw materials costs as well as strikes and bad weather. The ultimate sanction, of course, is to terminate the partnership. In the Prato area, this could happen if, for instance, an *impannatore* failed to pass orders back to the weaver who had supplied the fabric design that was being sold.

It seems clear that, for at least some value-added chains, a value-adding partnership is a viable and advantageous means of achieving the benefits of vertical integration. By observing the characteristics of and the processes followed by successful partnerships, executives can determine whether VAPs might pay off for their organizations. Business relationships premised on the need to achieve bargaining power may be more aggressively competitive than is in their best interest. Remember that the examples cited earlier – U.S. automobiles, Italian textiles, and drug distribution – all evolved from competitive, sometimes acrimonious, relationships.

The economic logic of the VAP is compelling. And at least for now, VAPs are part of the business landscape. Others should mind their feeding grounds and watering holes, for even giants have their vulnerabilities.

Reprint 88407

"The 'Businessman's Lunch,' sir? Perhaps if I could see some identification."

Collaborative Design

The manufacturing engineer asks, "What good is it if I can't make it?"

Manufacturing by Design

by DANIEL E. WHITNEY

In many large companies, design has become a bureaucratic tangle, a process confounded by fragmentation, overspecialization, power struggles, and delays. An engineering manager responsible for designing a single part at an automobile company told me recently that the design process mandates 350 steps–not 350 engineering calculations or experiments but 350 workups requiring 350 signatures. No wonder, he said, it takes five years to design a car; that's one signature every 3½ days.

It's not as if companies don't know better. According to General Motors executives, 70% of the cost of manufacturing truck transmissions is determined in the design stage. A study at Rolls-Royce reveals that design determines 80% of the final production costs of 2,000 components.[1] Obviously, establishing a product's design calls for crucial choices–about materials made or bought, about how parts will be assembled. When senior managers put most of their efforts into analyzing current production rather than product design, they are monitoring what accounts for only about a third of total manufacturing costs–the window dressing, not the window.

Moreover, better product design has shattered old expectations for improving cost through design or redesign. If managers used to think a 5% improvement was good, they now face competition that is reducing drastically the number of components and subassemblies for products and achieving a 50% or more reduction in direct cost of manufacture. And even greater reductions are coming, owing to new materials and materials-processing techniques. Direct labor, even lower cost labor, accounts for so little of the total picture that companies still focusing on this factor are misleading themselves not only about improving products but also about how foreign competitors have gained so much advantage.

In short, design is a strategic activity, whether by intention or by default. It influences flexibility of sales strategies, speed of field repair, and efficiency of manufacturing. It may well be responsible for the company's future viability. I want to focus not on the qualities of products but on development of the processes for making them.

Design is a strategic activity, by intention or by default.

Converting a concept into a complex, high-technology product is an involved procedure consisting of many steps of refinement. The initial idea never quite works as intended or performs as well as desired. So designers make many modifications,

Daniel E. Whitney is a section chief in the Robotics and Assembly Systems Division at the Charles Stark Draper Laboratory, Inc. in Cambridge, Massachusetts. He and his colleagues are active in research and industrial consulting, particularly with regard to the design and assembly of products. Mr. Whitney's article "Real Robots Do Need Jigs" appeared in HBR's May-June 1986 issue.

1. J. Corbett, "Design for Economic Manufacture," *Annals of C.I.R.P.*, vol. 35, no. 1, 1986, p. 93.

including increasingly subtle choices of materials, fasteners, coatings, adhesives, and electronic adjustments. Expensive analyses and experiments may be necessary to verify design choices.

In many cases, designers find that the options become more and more difficult; negotiations over technical issues, budgets, and schedules become intense. As the design evolves, the choices become interdependent, taking on the character of an interwoven, historical chain in which later decisions are conditioned by those made previously.

Imagine, then, that a production or manufacturing engineer enters such detailed negotiations late in the game and asks for changes. If the product designers accede to the requests, a large part of the design may simply unravel. Many difficult and pivotal choices will have been made for nothing. Where close calls went one way, they may now go another; new materials analyses and production experiments may be necessary.

Examples of failure abound. One research scientist I know, at a large chemical company, spent a year perfecting a new process–involving, among other things, gases–at laboratory scale. In the lab the process operated at atmospheric pressure. But when a production engineer was finally called in to scale up the process, he immediately asked for higher pressures. Atmospheric pressure is never used in production when gases are in play because maintaining it requires huge pipes, pumps, and tanks. Higher pressure reduces the volume of gases and permits the use of smaller equipment. Unfortunately, the researcher's process failed at elevated pressures, and he had to start over.

> If manufacturing joins the team too late, hard choices will have been made for nothing.

Or consider the manufacturer whose household appliance depended on close tolerances for proper operation. Edicts from the styling department prevented designs from achieving required tolerances; the designers wanted a particular shape and appearance and would not budge when they were apprised of the problems they caused to manufacturing. Nor was the machine designed in modules that could be tested before final assembly. The entire product was built from single parts on one long line. So each finished product had to be adjusted into operation–or taken apart after assembly to find out why it didn't work. No one who understood the problem had enough authority to solve it, and no one with enough authority understood the problem until it was too late. This company is no longer in business.

Finally, there was the weapon that depended for its function on an infrared detector, the first of many parts–lenses, mirrors, motors, power supplies, etc.–that were glued and soldered together into a compact unit. To save money, the purchasing department switched to a cheaper detector, which caused an increase in final test failures. Since the construction was glue and solder, bad units had to be scrapped. Someone then suggested a redesign of the unit with reversible fasteners to permit disassembly. But this time more reasonable voices prevailed. Reversible fasteners would have actually increased the weapon's cost and served no purpose other than to facilitate factory rework. Disassembly would not have been advisable because the unit was too complex for field repair. It was a single-use weapon–with a shelf life of five years and a useful life of ten seconds. It simply had to work the first time.

Manufacturers can avoid problems like this. Let's look at a success story. One company I know wanted to be able to respond in 24 hours to worldwide orders for its electronic products line–a large variety of features in small-order batches. Engineers decided to redesign the products in modules, with different features in each module. All the modules are plug compatible, electrically and mechanically. All versions of each module are identical on the outside where assembly machines handle them. The company can now make up an order for any set of features by selecting the correct modules and assembling them, all of this without any human intervention, from electronic order receipt to the boxing of final assemblies.

In another company, a high-pressure machine for supplying cutting oil to machine tools requires once-a-day cleaning. Designers recently reconfigured the machine so that normal cleanout and ordinary repairs can be accomplished without any tools, thus solving some bothersome union work-rule problems.

There are no guarantees, of course, but the experiences of these companies illustrate how design decisions should be integrated, informed, and balanced, and how important it is to involve manufacturing engineers, repair engineers, purchasing agents, and other knowledgeable people early in the process. The product designer asks, "What good is it if it doesn't work?" The salesperson asks, "What good is it if it doesn't sell?" The finance person asks, "What good is it if it isn't profitable?" The manufacturing engineer

2. David A. Garvin, "Competing on the Eight Dimensions of Quality," HBR November-December 1987, p. 101; and John R. Hauser and Don Clausing, "The House of Quality," HBR May-June 1988, p. 63.

asks, "What good is it if I can't make it?" The team's success is measured by how well these questions are answered.

The Design Team and Its Task

Multifunctional teams are currently the most effective way known to cut through barriers to good design. Teams can be surprisingly small–as small as 4 members, though 20 members is typical in large projects–and they usually include every specialty in the company. Top executives should make their support and interest clear. Various names have been given to this team approach, like "simultaneous engineering" and "concurrent design." Different companies emphasize different strengths within the team. In many Japanese companies, teams like this have been functioning for so long that most of the employees cannot remember another way to design a product.

Establishing the team is only the beginning, of course. Teams need a step-by-step procedure that disciplines the discussion and takes members through the decisions that crop up in virtually every design. In traditional design procedures, assembly is one of the last things considered. My experience suggests that assembly should be considered much earlier. Assembly is inherently integrative. Weaving it into the design process is a powerful way to raise the level of integration in all aspects of product design.

A design team's charter should be broad. Its chief functions include:

1. Determining the character of the product, to see what it is and thus what design and production methods are appropriate.

2. Subjecting the product to a product function analysis, so that all design decisions can be made with full knowledge of how the item is supposed to work and all team members understand it well enough to contribute optimally.

3. Carrying out a design-for-producibility-and-usability study to determine if these factors can be improved without impairing functioning.

4. Designing an assembly process appropriate to the product's particular character. This involves creating a suitable assembly sequence, identifying subassemblies, integrating quality control, and designing each part so that its quality is compatible with the assembly method.

5. Designing a factory system that fully involves workers in the production strategy, operates on minimal inventory, and is integrated with vendors' methods and capabilities.

The Product's Character. Clearly it is beyond the scope of this article to establish by what criteria one judges, develops, or revamps the features of products. Recently in these pages, David A. Garvin has analyzed eight fundamental dimensions of product quality; and John R. Hauser and Don Clausing have explored ways to communicate to design engineers the dimensions consumers want–in the engineers' own language.[2]

Character defines the criteria by which designers judge, develop, or revamp product features. I would only reiterate that manufacturing engineers and others should have something to say about how to ensure that the product is field repairable, how skilled users must be to employ it successfully, and whether marketability will be based on model variety or availability of future add-ons.

An essential by-product of involving manufacturing, marketing, purchasing, and other constituencies

"I guess I have to agree–in today's dicey investment climate, one does need all the help one can get."

Designing for Predictability: New MCAE Tools

A well-designed product is a predictable product. Managers particularly need to predict reliability, manufacturing costs, and manufacturability. In the past, engineers have dealt with these three issues only after engineering has completed the drawings, the near-final stage in the development cycle. But by then it may be too late. Moreover, when a product has hundreds, or even thousands of components, it's just not feasible to design, prototype, and redesign slightly different versions of each.

Two new, integrated, mechanical computer-aided engineering (MCAE) systems permit engineering teams to test before they build, so they can design for total quality with reliability, performance, and manufacturing costs in mind from the start. The first of these new systems model products not just physically but analytically: a designer can not only draw a component on the screen but prompt the computer to model and test the feasibility of its features, in the manner of a business planner using an electronic spreadsheet to play "what if" games. Engineers can vary assumptions about materials, speeds, loads, size, and other operating conditions. In this way, developers can both see the effects of hypothetical stresses and estimate product costs while making design decisions.

A company making internal-combustion engines, for example, may use an integrated MCAE system to design reliability and smoothness of operation into the counterbalance for the crankshaft. The optimal counterbalance keeps forces from the piston's up-and-down motion and centrifugal forces from the unbalanced crankshaft both low and constant – to reduce vibration. The system works as follows: a desktop workstation paints an image of a cylinder with all its operating parts on the screen. The engineer then selects values for counterbalance features like angle, thickness, diameter, etc., from menus. As choices are made, the system automatically computes the merits of the design, based on about 100 engineering equations, including compression ratio and stroke. So design variations are tried, evaluated, and discarded with near instantaneous response. This puts robustness and performance optimization into the very first counterbalance designs.

The second new system is an expert system which projects probable production costs for various part or assembly configurations and provides guidance as to their manufacturability. Another parts maker might use this system to project the manufacturability of and costs for, say, its stamped carburetor parts. Typically, the PC screen has several menus listing aspects of a part's design, like materials used and manufacturing process involved. Each menu holds progressive layers of possible choices. If the engineer selects metal from the list of materials, the system offers a choice of ferrous or nonferrous. Under ferrous metals, one can pick from carbon steel, stainless, cast iron, and so on.

There are automatic default values that the engineer might not normally specify, such as surface finish and carbon content. The system also draws on its own data base for manufacturing information like material density and base unit cost. Once the designer completes the menu sequence, the system produces an approximate part cost that includes materials, processing, and tooling expenditures.

MCAE systems don't eliminate prototyping entirely, but they can drastically reduce the number of trials and help engineers address reliability, performance, and cost problems early – and thus design for total quality from the very beginning.

–PHILIPPE VILLERS

Philippe Villers is president of Cognition Inc., a developer of MCAE systems in Billerica, Massachusetts. Previously, he founded Computervision Corporation in 1969 and Automatix Inc. in 1980.

in product conception, moreover, is that diverse team members become familiar enough with the product early in order to be able to incorporate the designers' goals and constraints in their own approaches. As designers talk with manufacturing or field-service reps, for example, they can make knowledgeable corrections. ("Why not make that part out of plastic? I know a low-cost source." "Because the temperature there is 1,000°; plastic will vaporize." "Oh.")

Product Function Analysis. This used to be the exclusive province of product designers. But now it is understood that to improve a product's robustness, to "design quality in" in Genichi Taguchi's good phrase, means thoroughly understanding a product's function in relation to production methods. Product designers and manufacturing engineers used to try to understand these relations by experience and intuition. Now they have software packages for modeling and designing components to guide them through process choices – software that would have been thought fantastic a generation ago. (See the insert, "Designing for Predictability: New MCAE Tools.")

Recently I worked on a product containing delicate spinning parts that had to be dynamically balanced to high tolerances. In the original design, partial disassembly of the rotating elements after balancing was necessary before the assembly could be finished, so the final product was rarely well balanced and required a lengthy adjustment procedure. Since total redesign was not feasible, the team analyzed the reassembly procedure solely as it pertained to balance and concluded that designers needed only to tighten various tolerances and reshape mating surfaces. Simple adjustments were then sufficient to restore balance in the finished product.

Another important goal of product function analysis is to reduce the number of parts in a product. The benefits extend to purchasing (fewer vendors and transactions), manufacturing (fewer operations, material handlings, and handlers), and field service (fewer repair parts).

When a company first brings discipline to its design process, reductions in parts count are usually easy to make because the old designs are so inefficient. After catching up, though, hard, creative work is necessary to cut the parts count further. One company I know saved several million dollars a year by eliminating just one subassembly part. The product had three operating states: low, medium, and high. Analysis showed that the actions of one part in the original design always followed or imitated the actions of two others. Designers eliminated the redundant part by slightly altering the shapes of the other two parts.

This change could never have been conceived, much less executed, if the designers hadn't had deep knowledge of the product and hadn't paid attention to the actions underlying its engineering.

Design for Producibility. Recently, a company bragged to a business newsweekly about saving a mere $250,000 by designing its bottles for a new line of cosmetics to fit existing machines for filling, labeling, and capping. This plan seems so obvious, and the savings were so small as compared with what is possible, that the celebration seemed misplaced. But it's a better outcome than I remember from my first job with a drug company. It spent a fortune to have a famous industrial designer create new bottles and caps for its line. They were triangular in cross section and teardrop shaped, and they would not fit either existing machines or any new ones we tried to design. The company eventually abandoned the bottles, along with the associated marketing campaign.

Obviously, nothing is more important to manufacturing strategy than designing for the production process. In the past, this has meant designing for manufacturing and assembly, and value engineering, which both strive to reduce costs. But now we have to go beyond these goals.

To take the last point first, value engineering aims chiefly to reduce manufacturing costs through astute choice of materials and methods for making parts. Does the design call for metal when a ceramic part will do? If metal, should we punch it or drill it? Value engineering usually comes into play after the design is finished, but the thoroughness we seek in design can be achieved only when decisions are made early.

Moreover, design for producibility differs from design for assembly, which typically considers parts one by one, simplifies them, combines some to reduce the parts count, or adds features like bevels around the rims of holes to make assembly easier. Valuable as it is, this process cannot achieve the most fundamental improvements because it considers the product as a collection of parts instead of something to satisfy larger goals, such as reducing costs over the product's entire life cycle.

> **Analyze the product to reduce the number of its parts; design the parts for jigless manufacture.**

Nippondenso's approach vividly illustrates how an overriding strategy can determine a product's parts and the production process. The Delco of Japan, Nippondenso builds such car products as generators, alternators, voltage regulators, radiators, and anti-skid brake systems. Toyota is its chief customer. Nippondenso has learned to live with daily orders for thousands of items in arbitrary model mixes and quantities.

The company's response to this challenge has several components:

- ☐ The combinatorial method of meeting model-mix production requirements.
- ☐ In-house development of manufacturing technology.
- ☐ Wherever possible, manufacturing methods that don't need jigs and fixtures.

The combinatorial method, carried out by marketing and engineering team members, divides a product into generic parts or subassemblies and identifies the necessary variations of each. The product is then designed to permit any combination of variations of these basic parts to go together physically and functionally. (If there are 6 basic parts and 3 varieties of each, for example, the company can build $3^6 = 729$ different models.) The in-house manufacturing team cooperates in designing the parts, so the manufacturing system can easily handle and make each variety of each part and product.

Jigless production is an important goal at this point, for obvious reasons. Materials handling, fabrication, and assembly processes usually employ jigs, fixtures, and tools to hold parts during processing and transport; the jigs and fixtures are usually designed specifically to fit each kind of part, to hold them securely. When production shifts to a different batch or model, old jigs and tools are removed and new ones installed. In mass-production environments, this changeover occurs about once a year.

In dynamic markets, however, or with just-in-time, batches are small, and shifts in production may occur hourly–even continually. It may be impossible to achieve a timely and economical batch-size-of-one production process if separate jigs are necessary for each model. Nippondenso's in-house manufacturing team responds to this problem by showing how to design the parts with common jigging features, so that one jig can hold all varieties, or by working with designers to make the product snap or otherwise hold itself together so that no clamping jigs are needed.

By cultivating an in-house team, Nippondenso also solves three difficult institutional problems. First, the company eliminates proprietary secrecy problems. Its own people are the only ones working on the design or with strategically crucial components. Second, equipment can be delivered without payment of a vendor's markup, thus reducing costs and making financial justification easier. Third, over the years the team has learned to accommodate itself intuitively to the company's design philosophy, and individual team members have learned how to contribute to it. Designers get to know each other too, creating many informal communication networks that greatly shorten the design process. Shorter design periods mean less lead time, a clear competitive advantage. (It is worth noting that many Japanese companies follow this practice of designing much of their automation in-house, while buying many product components from outside vendors. American companies usually take the opposite tack: they make many components and buy automation from vendors.)

Nippondenso uses combinatorial design and jigless manufacturing for making radiators (see the diagram). Tubes, fins, headers, and side plates comprise the core of the radiator. These four snap together, which obviates the need for jigs, and the complete core is oven soldered. The plastic tanks are crimped on. The crimp die can be adjusted to take any tank size while the next radiator is being put in the crimper, so radiators can be processed in any model order and in any quantity. When asked how much the factory cost, the project's chief engineer replied, "Strictly speaking, you have to include the cost of designing the product." A factory isn't just a factory, he implied. It is a carefully crafted fusion of a strategically designed product and the methods for making it.

How a Radiator Is Made–the Combinatorial, Jigless Method

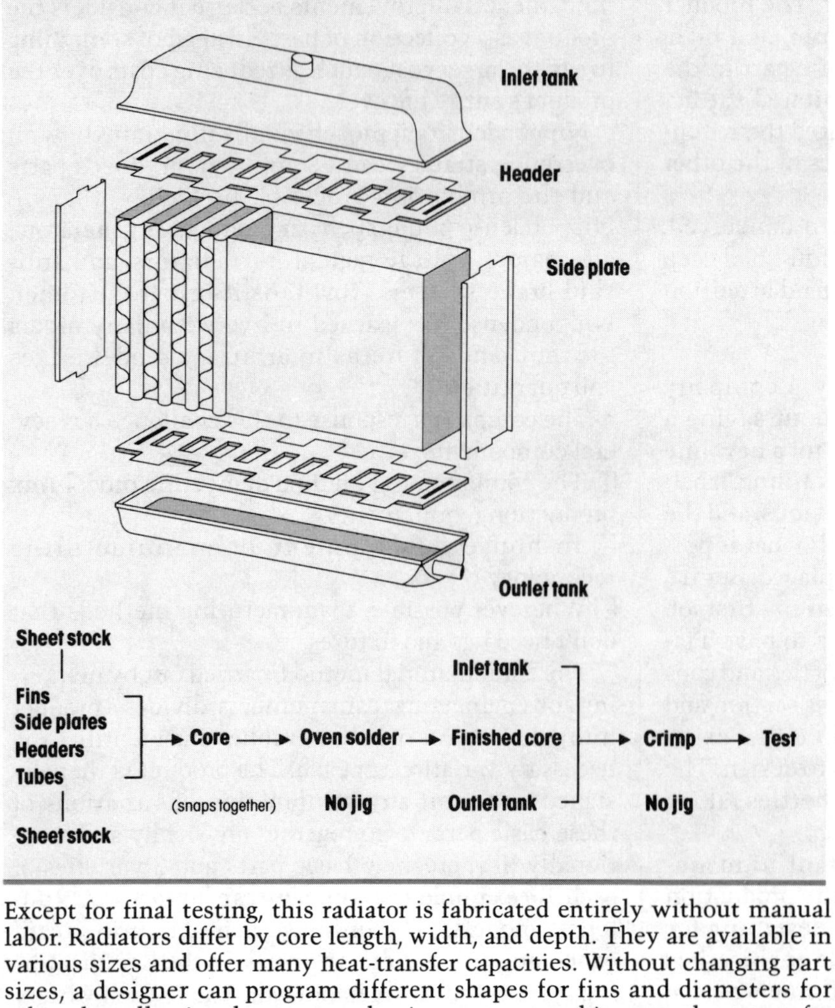

Except for final testing, this radiator is fabricated entirely without manual labor. Radiators differ by core length, width, and depth. They are available in various sizes and offer many heat-transfer capacities. Without changing part sizes, a designer can program different shapes for fins and diameters for tubes, thus allowing the same production system to achieve new heat-transfer capacities.

Without a guiding strategy, there is no way to tell what suggestions for improvement really support long-range goals. Some product-design techniques depend too much on rules, including rule-based systems stemming from expert systems. These are no substitute for experienced people. Volkswagen, for example, recently violated conventional ease-of-assembly rules to capture advantages the company would not otherwise have had.

In the company's remarkable Hall 54 facility in Wolfsburg, Germany, where Golfs and Jettas go through final assembly, robots or special machines perform about 25% of the final-stage steps. (Before Hall 54 began functioning, Volkswagen never did better than 5%.)[3]

To get this level of automation, VW production management asked to examine every part. It won from the board of directors a year-long delay in introducing the new models. Several significant departures from conventional automotive design practices resulted, the first involving front-end configuration. Usually, designers try to reduce the number of parts. But VW engineers determined that at a cost of one *extra* frame part the front of the car could be temporarily left open for installation of the engine by hydraulic arms in one straight, upward push. Installing the engine used to take a minute or longer and involved several workers. VW now does it unmanned in 26 seconds.

Another important decision concerned the lowly screw. Purchasing agents usually accept the rule that low-cost fasteners are a competitive edge. VW engineers convinced the purchasing department to pay an additional 18% for screws with cone-shaped tips that go more easily into holes, even if the sheet metal or plastic parts were misaligned. Machine and robot insertion of screws thus became practical. Just two years later, so many German companies had adopted cone-pointed screws that their price had dropped to that of ordinary flat-tip screws. For once, everyone from manufacturing to purchasing was happy.

Assembly Processes. Usually assembly sequence is looked at late in the design process when industrial engineers are trying to balance the assembly line. But the choice of assembly sequence and the identification of potential subassemblies can affect or be affected by–among other factors–product-testing options, market responsiveness, and factory-floor layout. Indeed, assembly-related activities with strategic implications include: subassemblies, assembly sequence, assembly method for each step, and integration of quality control.

Imagine a product with six parts. We can build it many ways, such as bottom up, top down, or from three subassemblies of two parts each. What determines the best way? A balance of many considerations: construction needs, like access to fasteners or lubrication points; ease of assembly (some sequences may include difficult part matings that risk damage to parts); quality control matters, like the operator's ability to make crucial tests or easily replace a faulty part; process reasons, like ability to hold pieces accurately for machine assembly; and, finally, production strategy advantages, like making subassemblies to stock that will be common to many models, or that permit assembly from commonly available parts.

Again, software now exists to help the designer with the formidable problem of listing all the possible assembly sequences–and there can be a lot, as many as 500 for an item as simple as an automobile rear axle. It would be impossible for a team to attack so complex a series of choices without a computer design aid to help, according to a preestablished hierarchy of goals like that just discussed–access to lubrication points, etc. Another virtue of this software is that it forces the team to specify choices systematically and reproducibly, for team members' own edification but also in a way that helps justify design and manufacturing choices to top management.

> Consider the sequence of assembly: Are you burying the parts you have to test?

Consider then, automatic transmissions, complex devices made up of gears, pistons, clutches, hydraulic valves, and electronic controls. Large transmission parts can scrape metal off smaller parts during assembly, and shavings can get into the control valves, causing the transmission to fail the final test or, worse, fail in the customer's car. Either failure is unacceptable and terribly expensive. It is essential to design assembly methods and test sequences to preempt them.

With respect to assembly machines and tooling, manufacturers should consider the following questions:

☐ Can the product be made by adding parts from one direction, or must it be turned over one or more times? Turnovers are wasted motion and costly in fixtures.

☐ As parts are added in a stack, will the location for each subsequent part drift unpredictably? If so, auto-

3. E.H. Hartwich, "Possibilities and Trends for the Application of Automated Handling and Assembly Systems in the Automotive Industry," International Congress for Metalworking and Automation, Hannover, West Germany, 1985, p. 126.

matic assembly machines will need expensive sensors to find the parts, or assembly will randomly fail, or parts will scrape on each other too hard.

☐ Is there space for tools and grippers? If not, automatic assembly or testing aren't options.

☐ If a manufacturing strategy based on subassemblies seems warranted, are the subassemblies designed so they do not fall apart during reorientation, handling, or transport?

There are clear advantages to combining consideration of these assembly procedures and/or quality control strategy with design. Designers who anticipate the assembly method can avoid pitfalls that would otherwise require redesign or create problems on the factory floor. They can also design better subassemblies to meet functional specifications – specifications that will be invaluable when the time comes to decide whether to take bids from outside vendors or make the part on the company's own lines, specifications that will determine how to test the subassembly before adding it to the final product.

> Manufacturable designs support automation, but they can also make it superfluous.

Designers concerned about assembly must ask:

☐ What is the best economic combination of machines and people to assemble a certain model-mix of parts for a product line (given each machine's or person's cost and time to do each operation, plus production-rate and economic-return targets)?

☐ How much time, money, production machinery, or in-process inventory can be saved if extra effort is put into design of the product, its fabrication and assembly processes, so that there are fewer quality control failures and product repairs? A process that yields only 80% successful assemblies on the first try may need 20% extra capacity and inventory – not to mention high-cost repair personnel – to meet the original production goals.

☐ Where in the assembly process should testing take place? Considerations include how costly and definitive the test is, whether later stages would hide flaws detectable earlier, and how much repaired or discarded assemblies would cost.

These are generic problems; they are hard to answer, and they too are stimulating the development of new software packages. This new software enhances the ability of manufacturing people to press their points in (often heated) debates about design. Hitherto, product designers, more accustomed to using computer modeling, have had somewhat of an upper hand.

Factory System Design. Many features of good product design presuppose that machines will do the assembly. But automation is not necessary to reap the benefits of strategic design. Indeed, sometimes good design makes automatic assembly unnecessary or uneconomic by making manual assembly so easy and reliable. Regardless of the level of automation, some people will still be involved in production processes, and their role is important to the success of manufacturing.

Kosuke Ikebuchi, general manager of the General Motors-Toyota joint venture, New United Motors Manufacturing Inc. (NUMMI), believes that success came to his plant only after careful analysis of the failures of the GM operation that had preceded it: low-quality parts from suppliers, an attitude that repair and rework were to be expected, high absenteeism resulting in poor workmanship, and damage to parts and vehicles caused by transport mechanisms.[4] The assembly line suffered from low efficiency because work methods were not standardized, people could not repair their own equipment, and equipment was underutilized. Excess inventory, caused by ineffective controls, was another problem. Work areas were crowded. Employees took too much time to respond to problems.

NUMMI's solutions focused on the Jidoka principle – quality comes first. According to NUMMI's factory system today, workers can stop the line if they spot a problem; the machinery itself can sense and warn of problems. Two well-known just-in-time methods of eliminating waste – the kanban system of production control and reductions in jig and fixture change times – are important to NUMMI's manufacturing operation.

But lots of other things also contribute to this plant's effectiveness: simplified job classifications, displays and signs showing just how to do each job and what to avoid, self-monitoring machines. NUMMI has obtained high-spirited involvement of the employees, first by choosing new hires for their

4. Kosuke Ikebuchi, unpublished remarks at the Future Role of Automated Manufacturing Conference, New York University, 1986.

Author's note: I am indebted to my colleagues James L. Nevins, Alexander C. Edsall, Thomas L. De Fazio, Richard E. Gustavson, Richard W. Metzinger, Jonathan M. Rourke, and Donald S. Seltzer for their contributions to this article. We have worked together for many years developing the ideas expressed here.

willingness to cooperate, then by training them thoroughly and involving them in decisions about how to improve the operations.

Design Means Business

The five tasks of design bring us back to the original point. Strategic product design is a total approach to doing business. It can mean changes in the pace of design, the identity of the participants, and the sequence of decisions. It forces managers, designers, and engineers to cross old organizational boundaries, and it reverses some old power relationships. It creates difficulties because it teases out incipient conflict, but it is rewarding precisely because disagreements surface early, when they can be resolved constructively and with mutual understanding of the outcome's rationale.

Strategic design is a continual process, so it makes sense to keep design teams in place until well after product launching when the same team can then tackle a new project. Design–it must be obvious by now–is a companywide activity. Top management involvement and commitment are essential. The effort has its costs, but the costs of not making the effort are greater.

Reprint 88412

GETTING THINGS DONE

Four ways to get design and manufacturing together

Organizing for Manufacturable Design

by James W. Dean, Jr. and Gerald I. Susman

Nowhere in a company is the need for coordination more acute than between the people who are responsible for product design and those responsible for manufacturing. As Daniel E. Whitney argued in these pages recently ("Manufacturing by Design," July-August 1988), most companies have operated for years in an environment where design and manufacturing communicate infrequently, if at all. In the worst instances, product designs were just thrown "over the wall": designers felt that their job was finished when designs were released and disappeared into manufacturing's domain; manufacturing engineers struggled to build products that were dropped in their laps.

Many companies have come to appreciate the disadvantages of this sequential approach to product development. Final designs emerging from engineering may be producible only at very high cost. While design expenditures *per se* may amount to only a small part of a product's total cost, design determines a huge proportion of producing, testing, and servicing costs. Forcing manufacturing to wait to begin its work until a design is released prolongs the development time of a product and may force a company to miss a market opportunity. Often a company is forced to play catch-up by implementing numerous engineering changes long after products have been introduced.

A first step: Buy designers software and give manufacturing a veto.

Effective manufacturers work from designs that have as few parts as possible, as many standard parts as possible, and that can be assembled by methods within manufacturing's capabilities. Barely manufacturable designs compromise product reliability and may preclude the use of robots or computer numerically controlled machines whose purchase was justified on the assumption that they could be kept busy.

Companies that have tried to design for manufacture merely by exhorting designers to create more producible designs, that is, without changing the basic organization of product development, have run into serious trouble. There are, after all, barriers to the integration of design and manufacturing. Often engineers in the two corporate functions have had different educations and share neither a common language nor compatible goals. Design engineers tend to be more focused on the product's performance or on its aesthetics; manufacturing engineers generally concentrate on plant efficiency.

In fact, designers generally enjoy higher status and pay than manufacturing engineers. Designers are considered something akin to creative artists and may be rewarded for ingenuity that has little to do with whether or not their designs can be turned into products cheaply and easily. Manufacturing people, whatever their background, often bear the stigma of less well-educated people, managers who have worked their way up from the factory floor.

The personnel from the two functions may also be located in different buildings, cities, or even countries, and there may be no real opportunity for them to establish rapport. Budgeting practices may exacerbate rivalries, as manufacturing engineers are often not funded to work on projects until the design is released, which is precisely when funding for design ends.

Having observed numerous manufacturing organizations, we've dis-

James W. Dean, Jr. is assistant professor of organizational behavior, and Gerald I. Susman is Robert and Judith Klein Professor of Management at the College of Business Administration of the Pennsylvania State University. Mr. Dean is associate director and Mr. Susman is director of the Center for the Management of Technological and Organizational Change there. Among Mr. Dean's publications is his book Deciding to Innovate *(Ballinger, 1987) and among Mr. Susman's, "People Policies for the New Machines" (HBR March-April 1987), written with Richard E. Walton.*

cerned several organizational approaches to designing for manufacture that go a long way toward overcoming these barriers. They range from manufacturing sign-off on designs, to combining product and process engineering into the same department. Some are more sweeping than others, but all are quite practical, having been used by a variety of companies. Fundamental to all of the approaches is basic change in the *structure* of the organization.

Manufacturing sign-off. In this approach, manufacturing engineers are given veto power over product designs, which cannot be released without manufacturing's approval, though in some cases, only its final approval. It is unlikely with this approach that an unproducible or barely producible design will reach the factory floor. But the approach's biggest drawback, clearly, is its heavy-handedness: it gives a club to manufacturing without providing for creative interchange between the two functions and does not allow manufacturing to begin its work until design's work is completed.

Companies using this approach seldom let designers grope blindly for what will satisfy manufacturing. Designers can use commercially available software to assess a product's producibility. Among the best known systems are Boothroyd and Dewhurst's Design for Assembly and Hitachi's Assemblability Evaluation Method. These and other systems calculate a producibility score for nearly any product, based on the number of its parts, the number of its standardized parts, the simplicity of couplers, the motions involved in its assembly, and so on. Some programs can even generate bar charts that demonstrate the contributions of subassemblies to manufacturing cost and time.

The manufacturing sign-off is relatively simple to manage and depends little on interpersonal skills of engineers on either side of the wall. Designers' use of expert-system software permits smaller companies to benefit from the accumulated knowledge of more advanced companies and allows manufacturing engineers to focus on their principal task, namely, design of the process.

Incidentally, a number of companies have created customized software for their design engineers. Customized packages aim to give designers information about the spe-

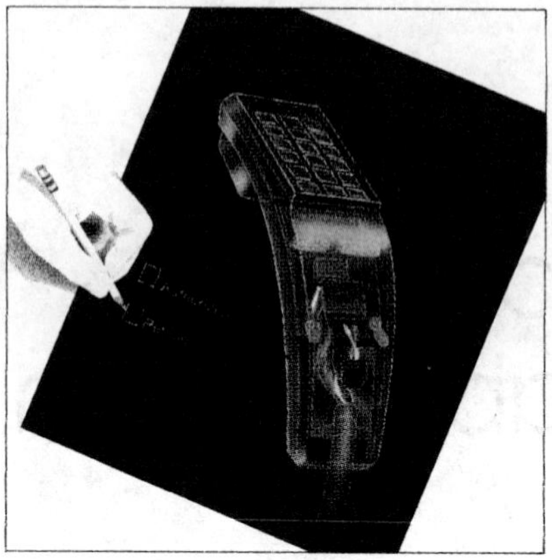

cific constraints of the production site. Other companies use less sophisticated (and less expensive) paper versions of the same idea, that is, listings of preferred components, standard production routings, and so forth. These documents are typically distributed to the design group, so they can try to make their designs manufacturable before manufacturing approval is solicited.

One appliance manufacturer we know, with several billion dollars in annual sales, has its product designers use a number of self-checking software modules to monitor such qualities as moldability (for plastics), formability (for metal drawings), and ease of automated assembly. The company wants products to achieve acceptable scores on manufacturability before manufacturing even sees the designs. This allows designers to see these programs as a tool to help them, rather than something that manufacturing uses to frustrate them. Approval must be granted by a producibility-feasibility group, however, at as many as ten phases in the design process. If approval is withheld, the design cannot pass to the final phases, when designers would ordinarily order equipment to build prototypes.

Senior managers at this appliance company believe that their software programs have educated designers and that they've helped the company make major strides in cost and quality. Again, however, the system's greatest weakness is the lack of day-to-day communication between designers and manufacturing people. As a manager who helped to design the current system put it, designers still "work in a vacuum." He wishes there were more interaction "when the paper is clean."

The company has considered bringing design and manufacturing into the same building, though the cost is currently prohibitive. Plans for shifting responsibility for building prototypes from design to manufacturing are underway. Executives are also contemplating letting manufacturing contract for toolmaking by outside vendors before the design is finalized so vendors can be "partners" in the process. However, such an approach may lead to complex (and possibly costly) negotiations with these vendors whenever a design is changed, and managers in finance are resisting it.

The integrator. Integrators work with designers on producibility issues, serving as liaisons to the manufacturing group. Naturally, such a role requires individuals who can keep design and manufacturing perspectives in balance. An integrator who leans too heavily toward manufacturing will lose credibility with designers, and someone who leans too heavily toward design will simply not get the job done.

> The integrator usually becomes the resident expert.

Given the way engineers are currently educated and promoted, integrator candidates may be hard to find. Manufacturing and design engineers are the products of separate and distinct degree programs. And once working for a company, they tend to be promoted within their respective hierarchies. There is usually little opportunity to broaden their provincial outlook to include the other group's concerns.

The integrator approach has been used by the electronics division of a multibillion dollar company we know, a company that sells advanced avionics and communications equipment to the military. Now that

the Defense Department insists that its contractors make provision for the producibility, serviceability, and maintainability of weapons systems, the division must pursue design for manufacture even more carefully. Moreover, since it recently spent tens of millions of dollars on an advanced electronics assembly plant, the company wants to see its products designed to take advantage of this new facility.

Company executives originally hired as integrator an industrial engineer with a background in design. He began by preparing a manual for the design engineers to use and by securing top-management support for manufacturing sign-off on product designs. The manual, planned as a collaboration between design and industrial engineering, was to be divided into five volumes, covering such things as lists of standard components and circuit-board mountings to the more obscure cost implications of producibility mistakes in complex, low-rate products.

Company executives also assigned a test engineer to make sure components were manufactured in a way that allowed for systematic inspection, and they issued a directive that the producibility coordinators were to approve all designs before release. As one supervisor put it, "The preachers became auditors."

This division has already seen substantial payoffs from these efforts. Boards on which automated component placement techniques can be used have increased from 40% to about 55%; for newly designed boards, almost 90%. Plans include designating a third producibility coordinator, as well as hiring a number of scientists to develop new production methods for the advanced designs that it will soon have to produce. The company also plans to involve its subcontractors in future efforts to improve producibility.

Clearly, the integrator approach is reasonably flexible. A single individual (or a small team) can easily keep track of new capabilities in manufacturing. Manufacturing engineers don't have to become more knowledgeable about design or designers become more expert in manufacturing. Rather, the approach develops an "expert" in producibility, who can become the focal point for companywide efforts.

There are some disadvantages, however. One problem is the downside of the integrator's virtue, what one integrator we spoke with called the "guru syndrome." Since the integrator is there to worry about producibility, no one else does. The integrator approach makes the organization very dependent on one (or only a few) individuals. It does not facilitate simultaneous engineering—that is, manufacturing cannot begin its work before design's work has been completed.

Cross-functional teams. Another step away from the traditional approach is cross-functional teams. At a minimum these consist of a designer and a manufacturing engineer, who work together throughout the whole process. The team meets regularly or may even be located in the same office.

This is the first approach that facilitates simultaneous engineering: the manufacturing engineer becomes familiar with the design well before it is released and may even have had a hand in creating it. The manufacturing process can be partially if not completely planned before the design is finalized.

Perhaps the best example we know of the cross-functional team is a process-control company with annual sales of about $100 million. It became interested in producibility about two years ago, when, like the military contractor we just discussed, it began to utilize autoinsertion equipment for circuit-board production. The company was also discouraged by the failure of one of its recent products whose poor quality was attributed to a barely producible design.

Initially, individuals from manufacturing engineering and quality assurance were formally assigned to the design team, and from the first day of a new product program, the entire group met once or twice a week. A program manager, who acted as a mediator, determined the meeting schedule. Eventually people from test engineering, purchasing, and marketing joined the team.

The new approach created some frictions. Designers felt that quality assurance and manufacturing were preempting them and wondered why the company didn't trust them to create good designs independently. They also felt manufacturing's demands were often unrealistic, particularly concerning clearances among

Cross-functional teams mean collaboration from the start.

the various components. They were upset that the new system undermined their creativity.

These sentiments have all dissipated over time, largely because the company took the trouble to calculate the relative cost of various designs at the circuit-board level and proved beyond doubt the importance of using autoinsertable components. Costs are so much lower now with automated assembly that the team recently set a goal of using autoinsertable components 97% of the time. Incidentally, while conflicts between design and manufacturing have generally been settled in team meetings, final authority over design rests with the engineering manager.

The team approach has also been used successfully by a multibillion dollar aerospace company we've studied that produces for both military and commercial markets. The company wanted to decrease its development time for major products from 36 to 24 months, to reduce the number of engineering changes necessary to meet production standards, and to fully utilize its advanced manufacturing capabilities.

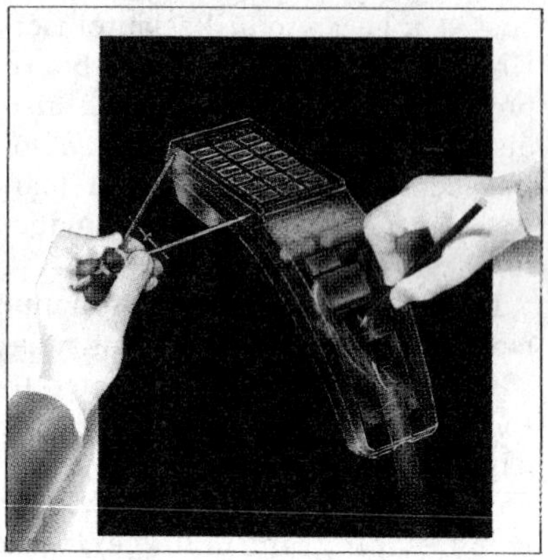

The producibility effort began in 1983 with a series of seminars for designers. By 1985, the company was ready to experiment with what it called product centers. It formed design teams with representatives from engineering, manufacturing, quality assurance, and product support (documentation, etc.), with either design or manufacturing providing leadership.

The product-center concept has continued to evolve. There are six now working on a single big project, organized around major subcomponents. Each product center has a manager and deputy manager and, whenever possible, people working on the project are located in the same room. Manufacturing begins process planning as soon as a week or two after design begins its work. Traditionally, manufacturing engineers would not begin to work until designs were released; today they *finish* almost simultaneously with design release.

The possibility of simultaneous engineering is one benefit of the team approach. Also, the frequent interaction involved (especially when product and process designers are located in the same place) permits people from the two functions to educate one another, thus enhancing capability for future efforts. Perhaps the most attractive feature of this team approach is the way it substitutes collaboration for auditing. Yet team members report through separate hierarchies, which helps them to hold to their respective missions. There are tensions, but these often stimulate greater creativity.

The team concept is expensive because it means assigning people to a development effort not only during the time when their special expertise is crucial but before and after as well. The approach requires members to gain broad expertise in producibility, since there is no longer a single producibility expert, and it demands excellent interpersonal skills. There are probably some engineers on both sides of the wall who would do more harm than good on such teams. Finally, the team approach runs the risk associated with any method that allows simultaneous engineering: process planning done before the final design is released may have to be scrapped as the design evolves.

The product-process design department. Our fourth approach to design for producibility involves the greatest degree of structural change. It entails creating a single department responsible for both product and process both. A number of variations of this approach are possible, including:

- □ A senior manager with responsibility for both product and process design, but separate subunits for each function.
- □ Product and process engineers combined into a single department with one manager having responsibility for both groups.
- □ One department composed of product-process engineers, that is, individuals with responsibility for both aspects of the design, a rarely found ideal, since very few people have the skills necessary to straddle both worlds.

One $100 million defense technology company we know learned the virtues of a product-process department the hard way. At first it subjected products to nonbinding manufacturability reviews; this worked fairly well until the value engineering group came up with an idea for reducing the cost of a critical component by substituting a die casting for a machined blank. Manufacturing liked the concept but contended that vendor-supplied components would not perform as promised without costly rework. The value engineers dismissed manufacturing's objections, and manufacturing didn't press the issue.

Unfortunately, the protests were justified. Reprocessing increased the cost of the component by almost 100%, as much as $300,000 over the life of the program. Ironically, the company had already intended to move to the product-process department approach and had even begun giving the head of value engineering the added responsibility of managing the manufacturing engineering department. As it happened, unfortunately, this person was not comfortable with the manufacturing group and spent very little time there.

> A one-department approach puts a premium on technical and interpersonal skills.

Now the company is operating with an interim structure. In the absence of an organization that supports manufacturable designs, producibility depends heavily on the personal inclinations of the designers. The company's computer-aided design system is networked between design and manufacturing, and designers can discuss the implications of designs with manufacturing while both view the drawings on their screens. Unfortunately, this happens far less often than management would like.

An automobile components company, a subsidiary of one of the big three, has the second type of product-process department; design and process engineers report to the same manager at the first level. Product-process teams share software systems and offices, and members are encouraged to decide on a design among themselves. If they don't, their manager resolves the issue. The company adopted this approach in order to reduce development time and cost, and these goals have been achieved, in addition to a reduction in engineering changes after products are in production.

The one-department approach permits simultaneous engineering and leads to mutual education through day-to-day contact. It also places a high premium on the technical and interpersonal skills of department members. The department head must strike a balance between engineering functions and bring a great deal of expertise to the table.

The one-department approach creates the greatest degree of structural change of the four approaches. It may also create the greatest resistance, as people are torn from the comfortable surroundings of their professional and departmental loyalties. Perhaps the greatest danger of the one-department approach is that, as the design and manufacturing communities work more closely together, they may find it too easy to compromise on designs that are merely acceptable to both sides–rather than insisting on functional excellence from their own disciplinary standpoint.

It is clear from our examples that a number of different approaches can lead to producible designs. How can managers choose among the alternatives? The approaches range in impact on the organization from manufacturing sign-off (relatively low) to the product-process department (quite high). Higher impact approaches allow for collaboration and simultaneous engineering, while lower impact approaches do not. On the other hand, higher impact approaches place much greater demands on people, in terms of absorbing change and developing new skills, both technical and interpersonal.

Companies that enjoy substantial freedom in product and process design, and that have the organizational ability to rapidly absorb change, are well positioned to take advantage of the higher impact approaches. Such companies can profit

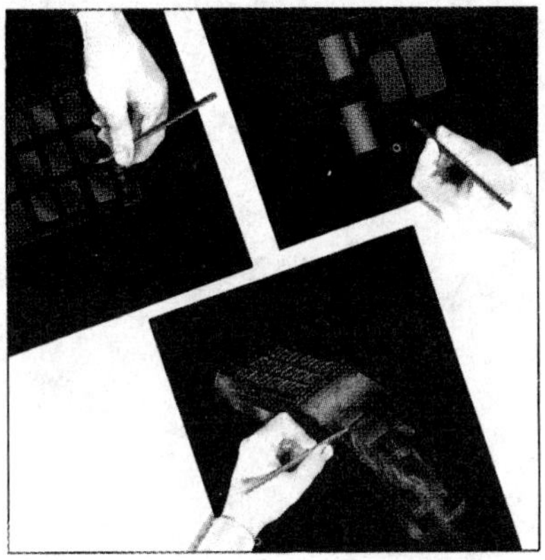

from the collaboration and simultaneous engineering that such methods afford.

Companies whose products and/or processes are relatively fixed, or whose capacity for absorbing change is limited, would be better advised to begin with manufacturing sign-off or the integrator approach. When products or processes are fixed, there is little to be gained by the intense interaction that the higher impact approaches feature. And organizations slow to change would take so long to implement the higher impact designs that they would get greater benefit by using one of the simpler approaches. Of course, such companies could always move to higher impact designs at a later date.

These approaches are not cast in concrete or meant to seem exclusive of one another. Structures for organizations ought to accommodate the messy dilemmas managers face, so each organization should customize its own approach, using the four identified here as building blocks. Whatever the chosen method, change in organization structure will be necessary in a program to achieve manufacturable design.

Reprint 89103

• CUSTOM PUBLICATIONS • BOOKS • ARTICLES • VIDEOS • CUSTOM PUBLICATIONS • BOOKS • ARTICLES • VIDEOS • BOOKS •

READ THE FINE PRINT

REPRINTS
Telephone: 617-495-6192
Fax: 617-495-6985

Current and past articles are available, as is an annually updated index. Discounts apply to large-quantity purchases.

Please send orders to
HBR Reprints
Harvard Business School
Publishing Division
Boston, MA 02163.

HOW CAN *HARVARD BUSINESS REVIEW* ARTICLES WORK FOR YOU?

For years, we've printed a microscopically small notice on the editorial credits page of the *Harvard Business Review* alerting our readers to the availability of *HBR* articles.

Now we invite you to take a closer look at some of the many ways you can put this hard-working business tool to work for you.

IN THE CORPORATE CLASSROOM.

There's no more effective, or cost-effective, way to supplement your corporate training programs than in-depth, incisive *HBR* articles.

Affordable and accessible, it's no wonder hundreds of companies and consulting organizations use *HBR* articles as a centerpiece for management training.

IN-BOX INNOVATION.

Where do your company's movers and shakers get their big ideas? Many find the inspiration for innovation in the pages of *HBR*. They then share the wealth and spread the word by distributing *HBR* articles to company colleagues.

IN MARKETING AND SALES SUPPORT.

HBR articles are a substantive leave-behind to your sales calls. And they can add credibility to your direct mail campaigns. They demonstrate that your company is on the leading edge of business thinking.

CREATE CUSTOM ARTICLES.

If you want to pack even greater power in your punch, personalize *HBR* articles with your company's name or logo. And get the added benefit of putting your organization's name before your customers.

AND THERE ARE 500 MORE REASONS IN THE *HBR CATALOG.*

In all, the *Harvard Business Review Catalog* lists articles on over 500 different subjects. Plus, you'll find books and videos on subjects you need to know.

The catalog is yours for just $8.00. To order *HBR* articles or the *HBR Catalog* (No. 21019), call 617-495-6192. Please mention telephone order code 025A when placing your order. Or FAX us at 617-495-6985.

And start putting *HBR* articles to work for you.

Harvard Business School Publications

VE RI TAS

Call 617-495-6192 to order the *HBR Catalog.*

(Prices and terms subject to change.)

• BOOKS • CUSTOM PUBLICATIONS • ARTICLES • VIDEOS • CUSTOM PUBLICATIONS • BOOKS • ARTICLES • VIDEOS • BOOKS •